This Book Belongs To

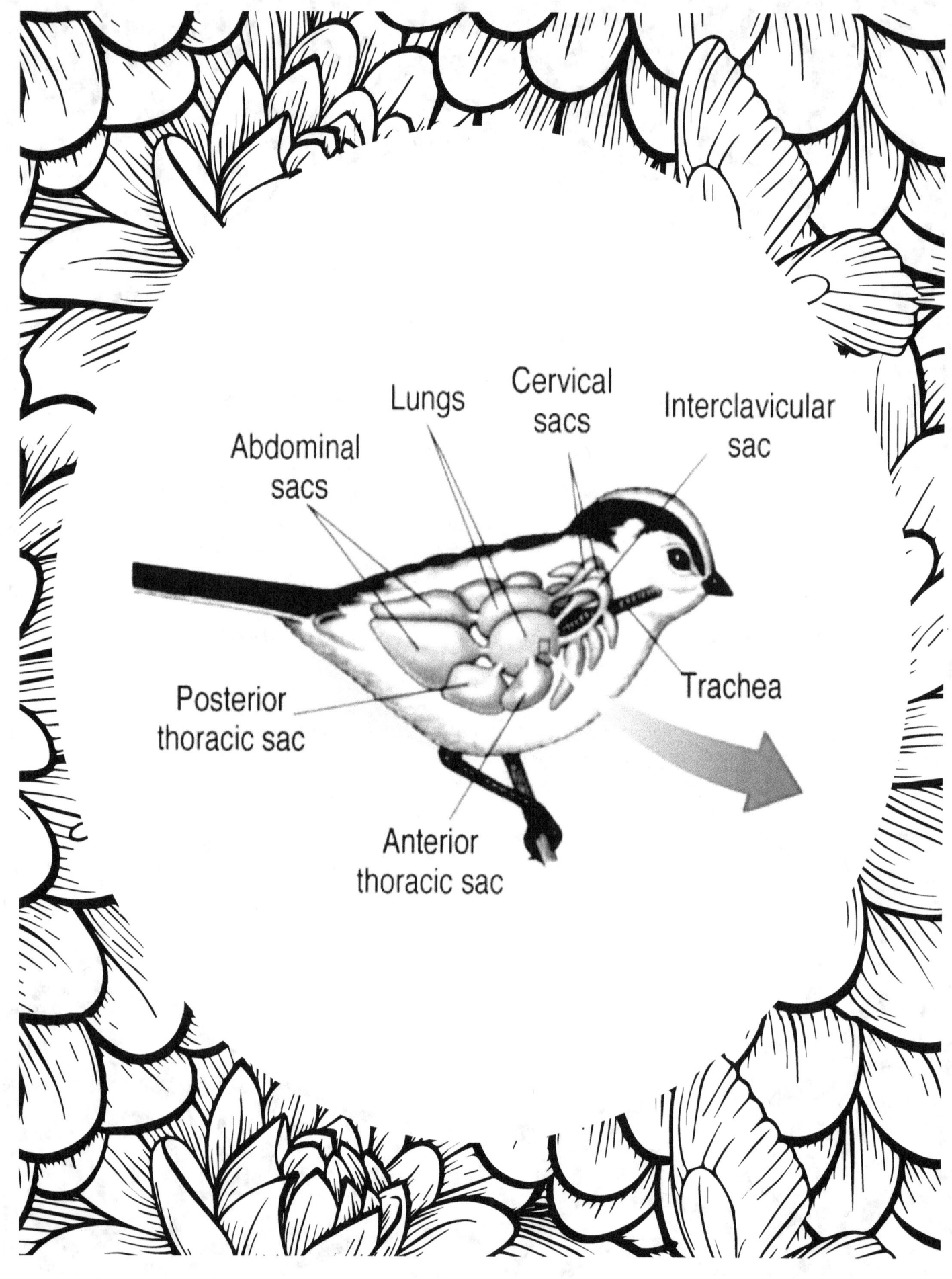

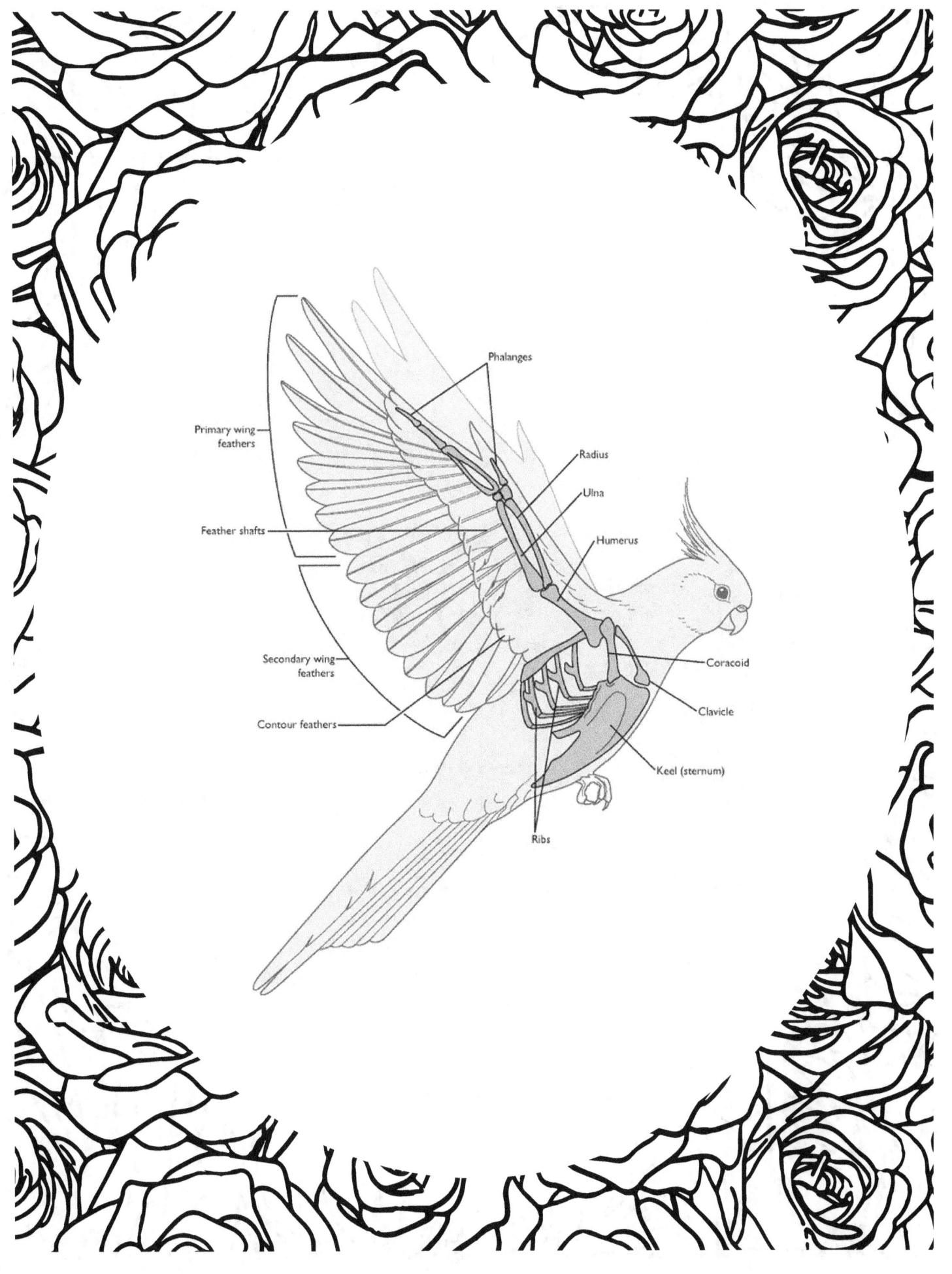

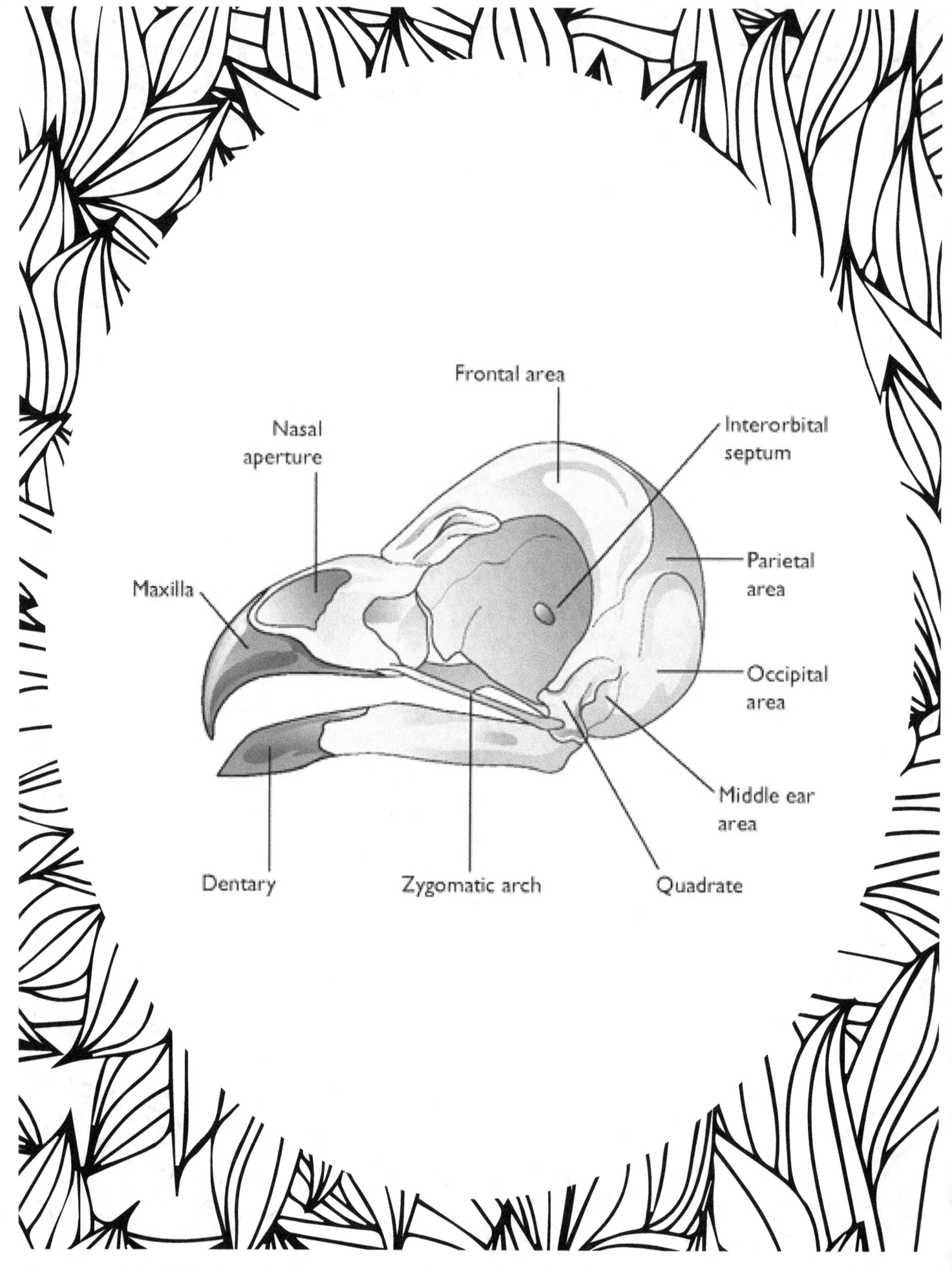

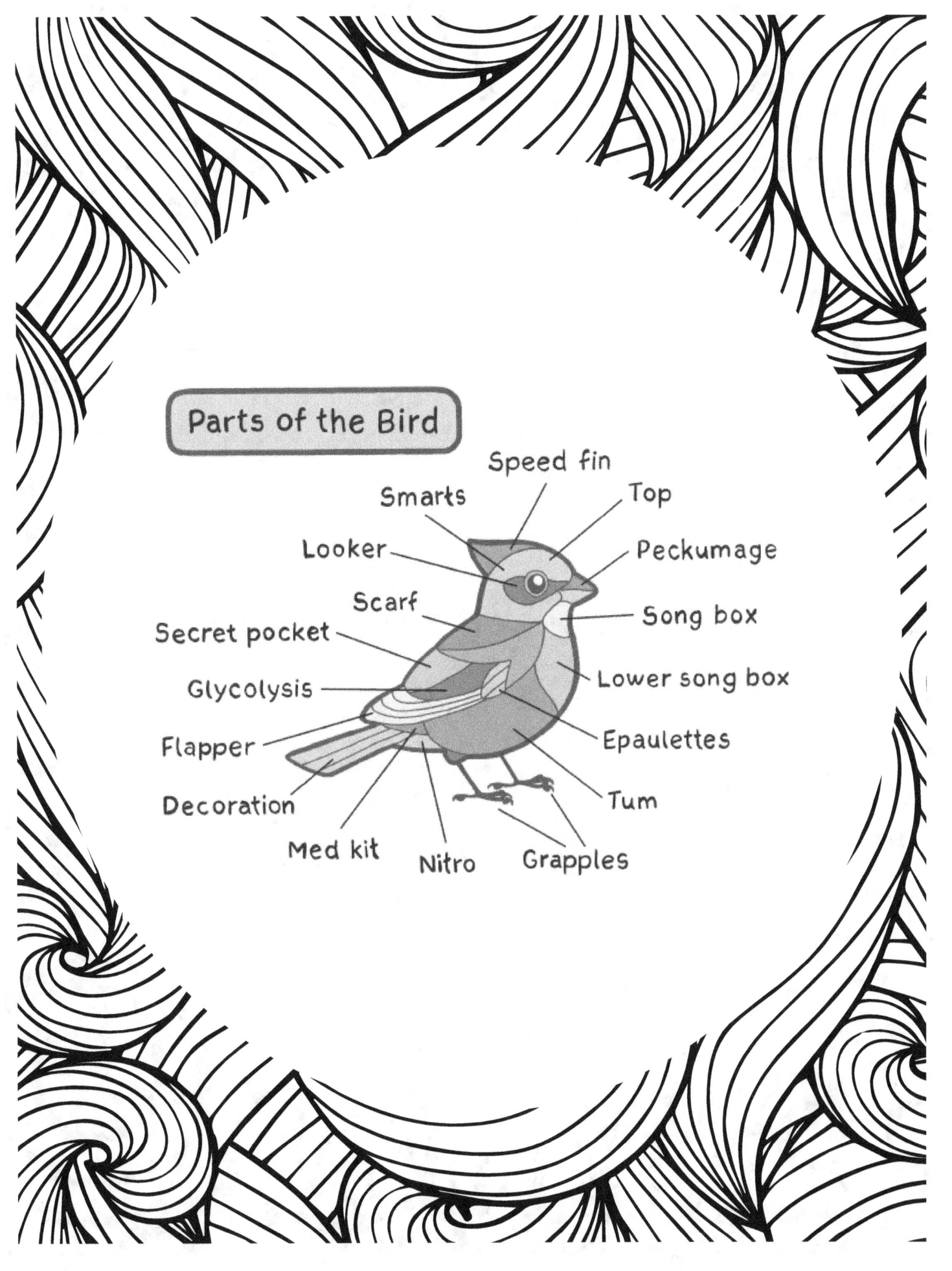

Parts of the Bird

Eyebrow or Supercilium

Bill

Chin

Throat

Breast

Alula

Belly

Primary-coverts

Flanks

Tibia

Tarsus

Claw

Toe

Abdomen

Undertail-coverts

Collar

Back

Mantle

Scapulars

Lesser coverts

Secondary-coverts

Secondaries

Primaries

Tail

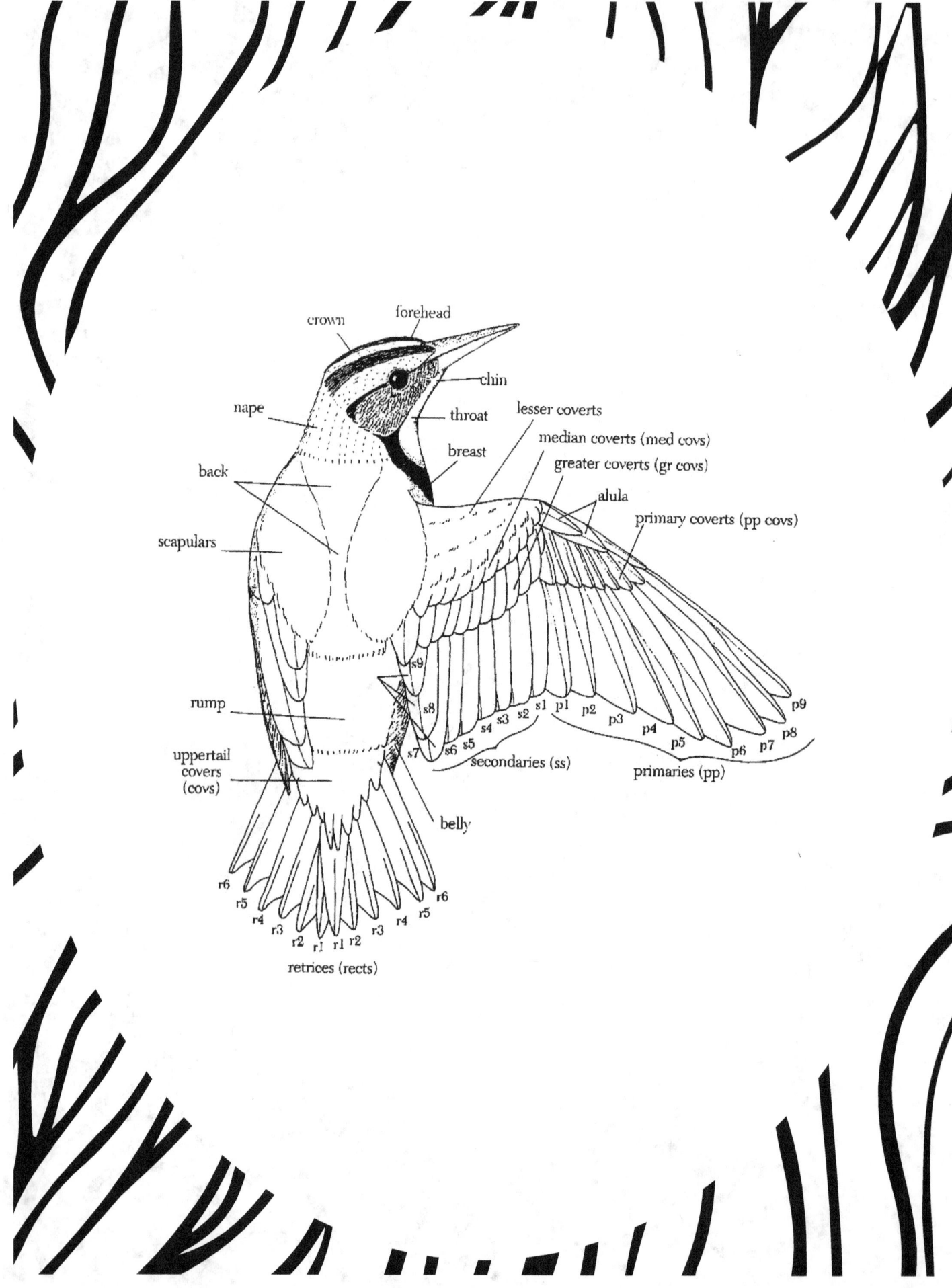

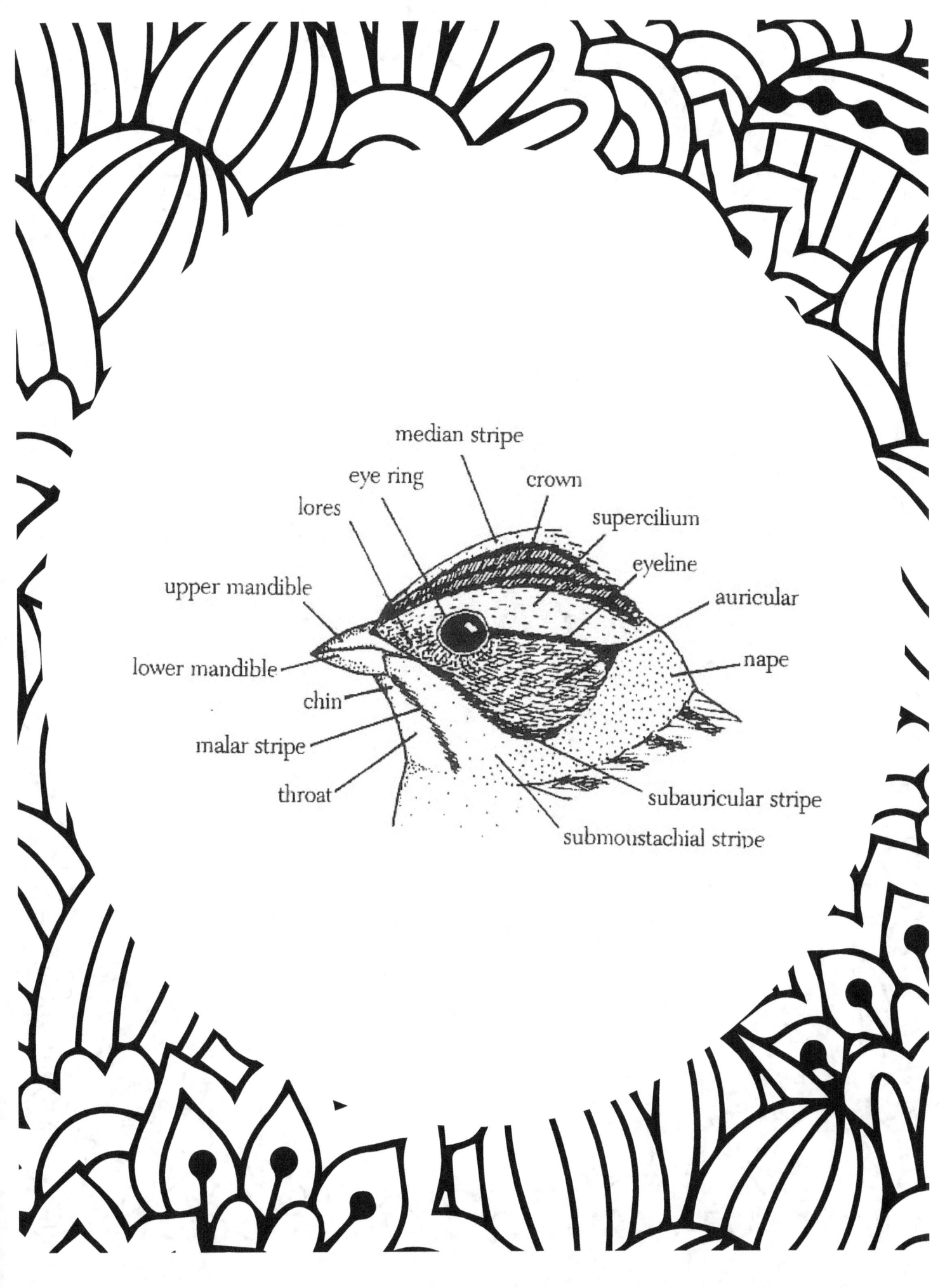

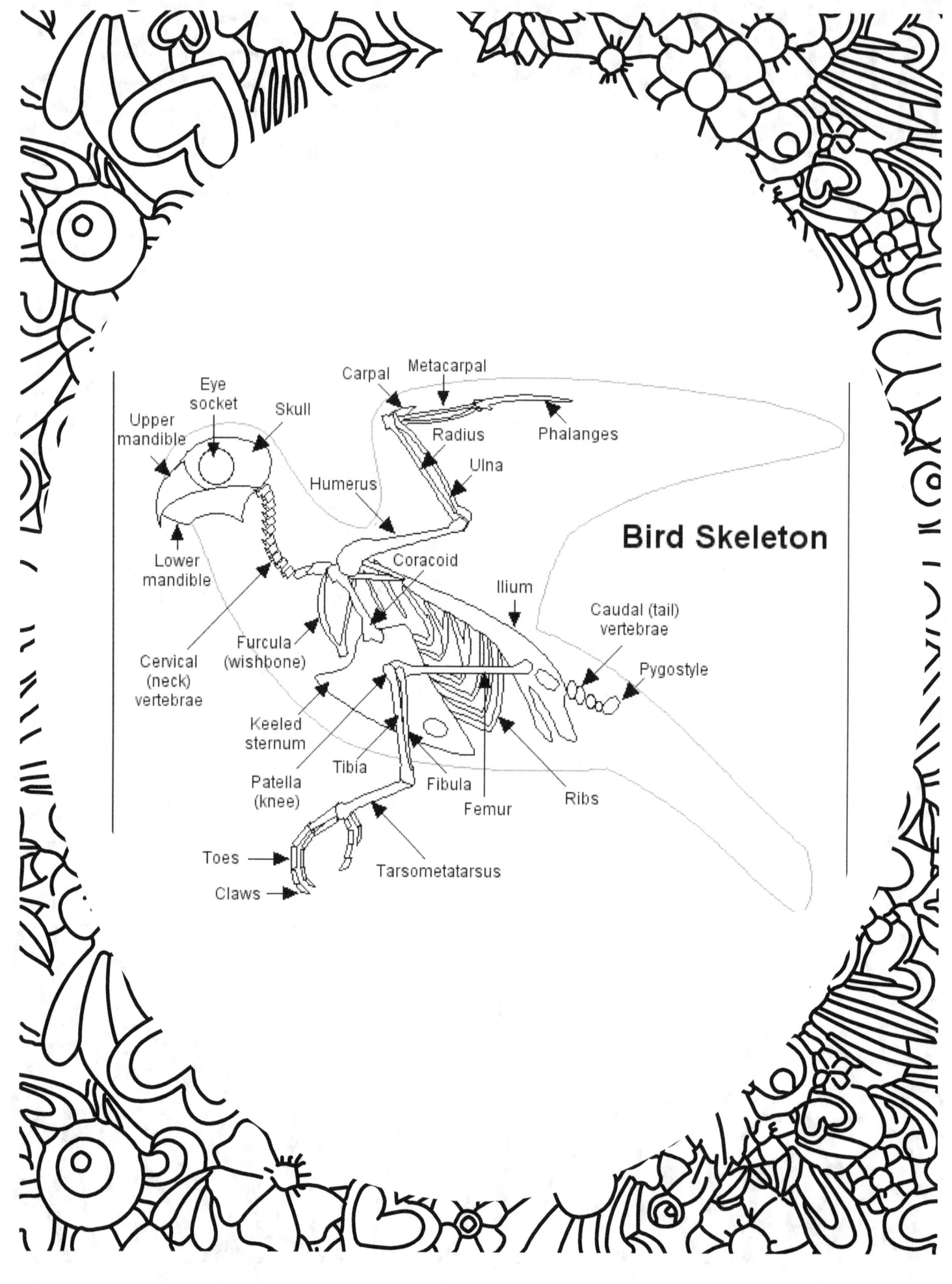

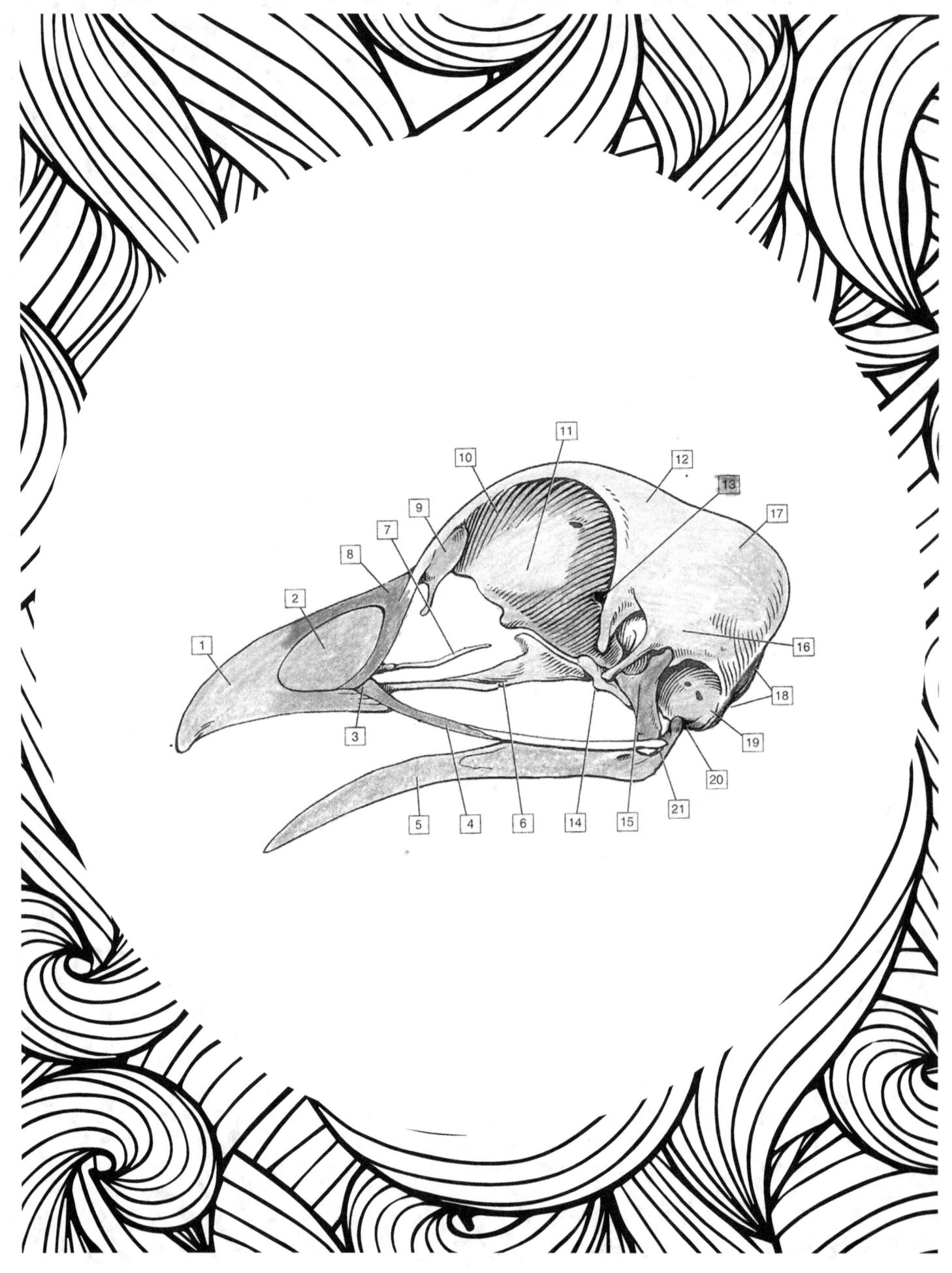

nape

scapulars

lesser coverts

median coverts

greater coverts

Marginal coverts Tiny feathers at bend of wing covering bony leading edge of "hand."

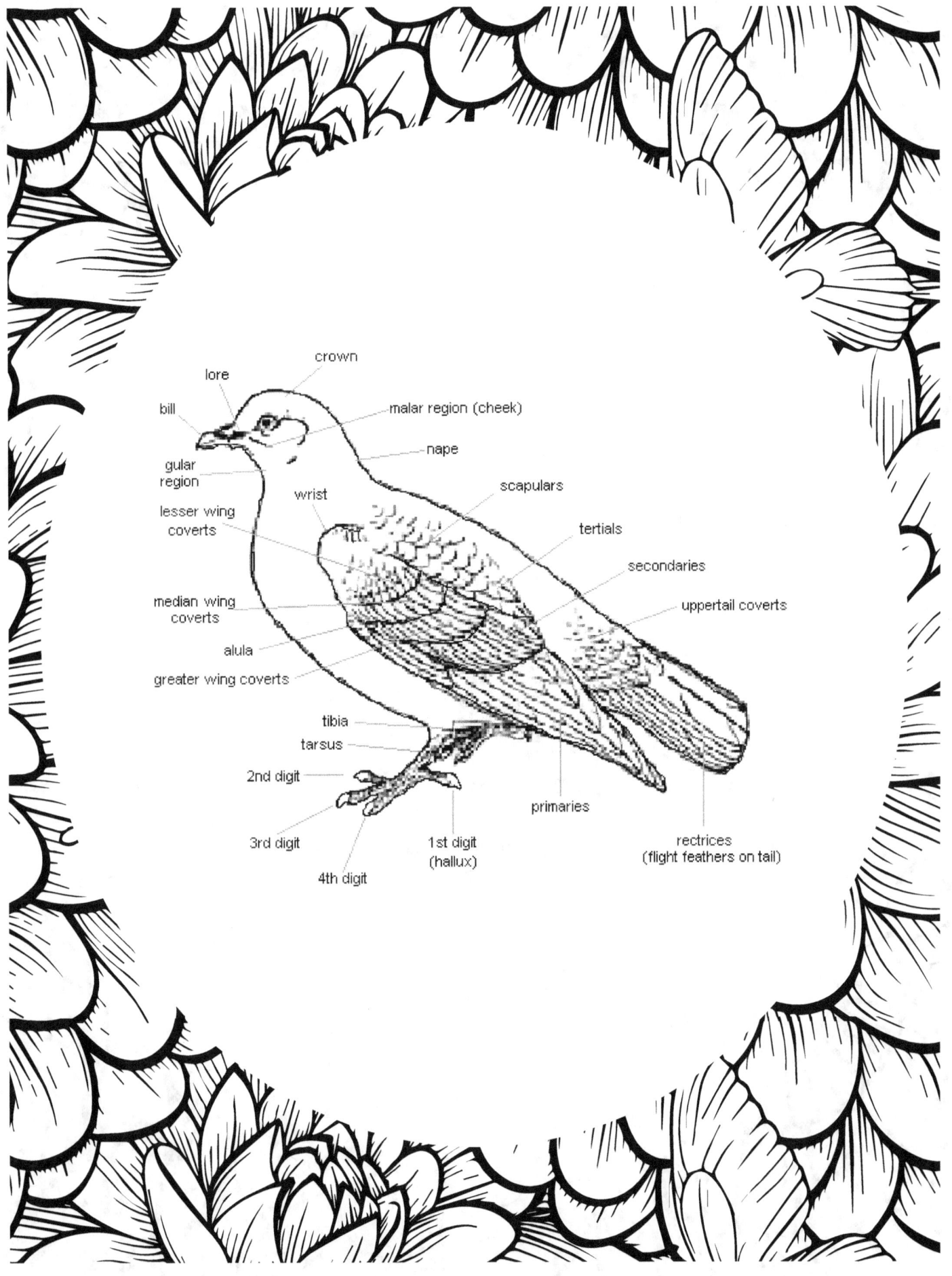

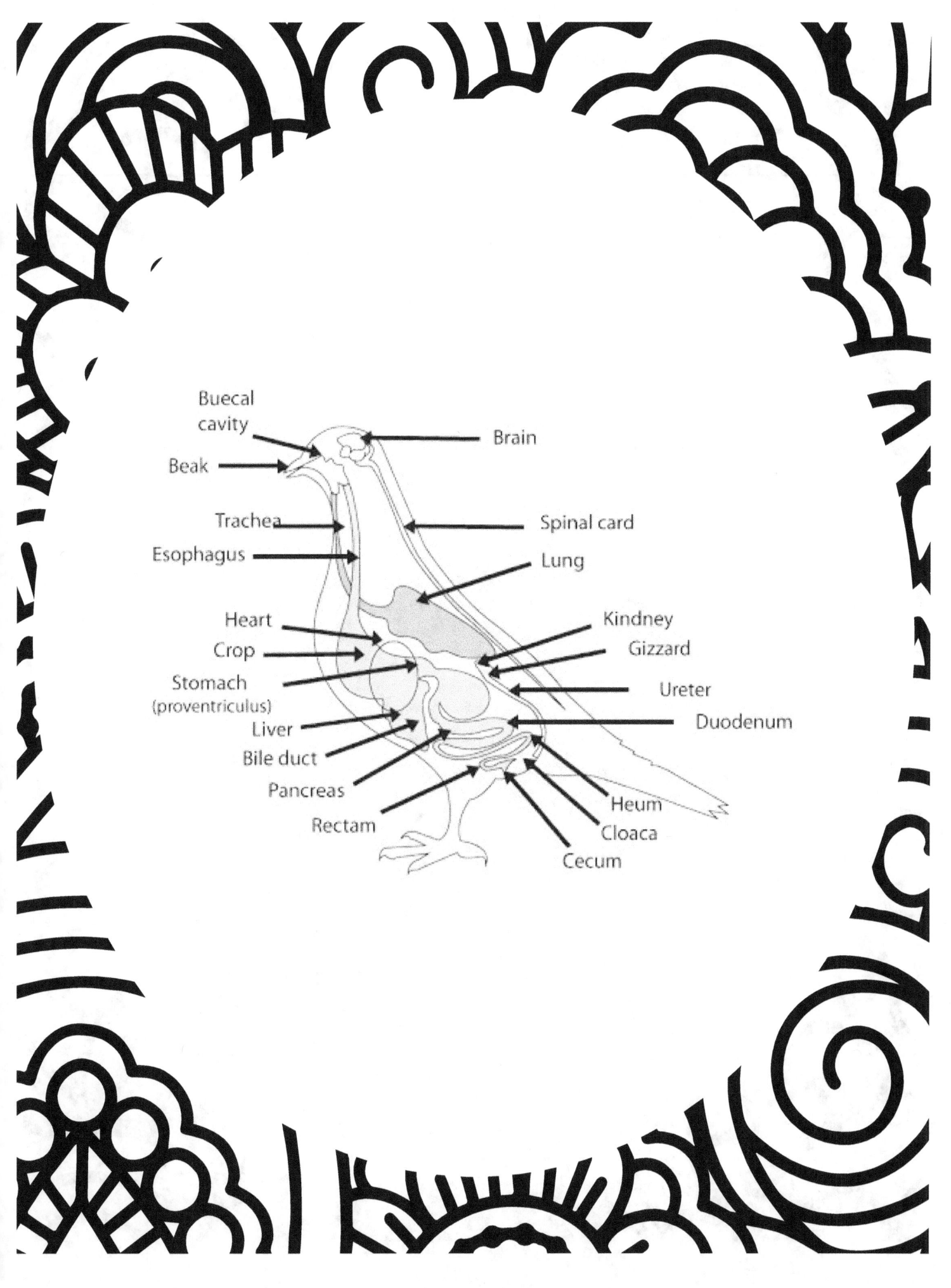

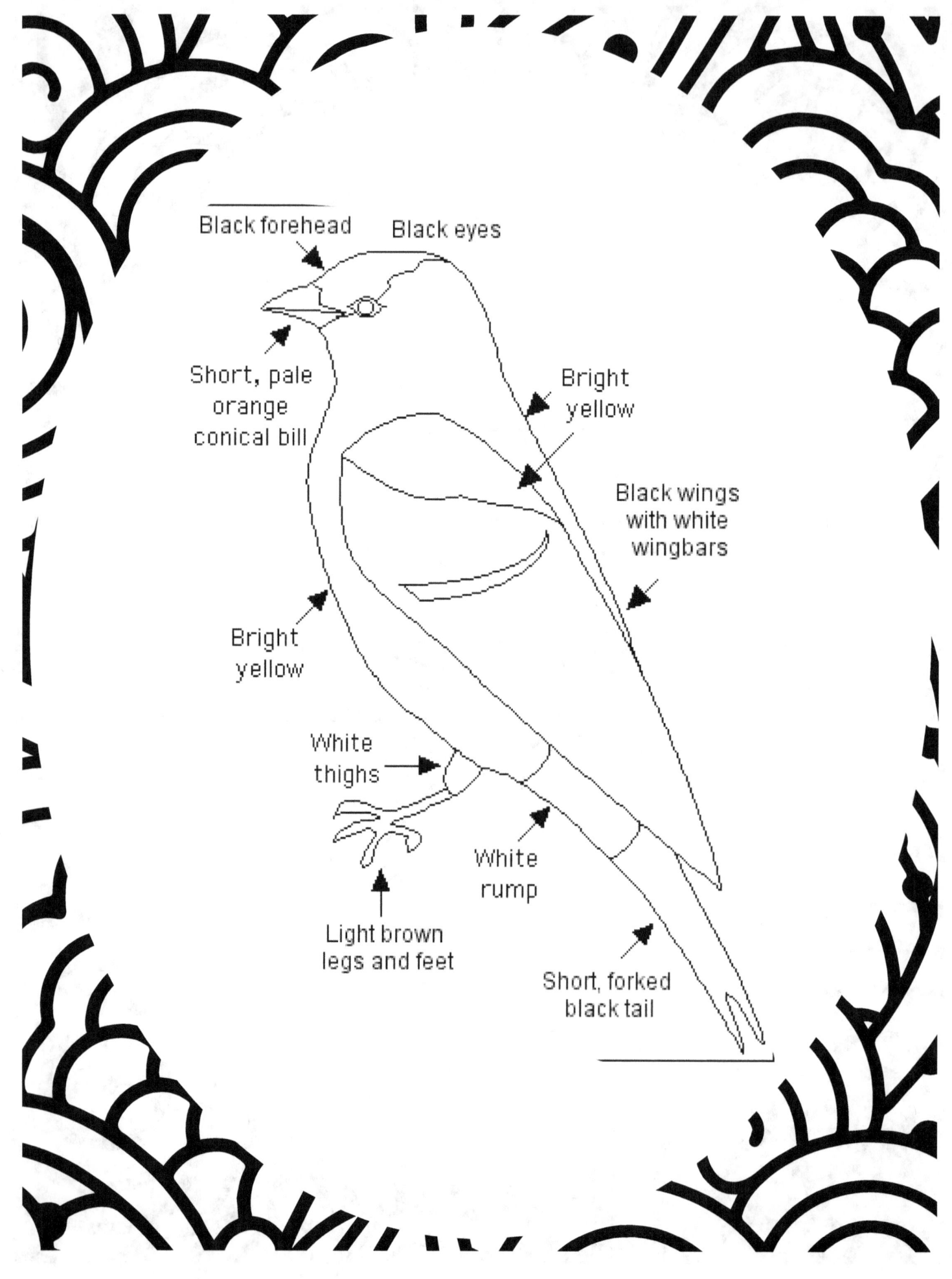

Black
bill
with
light
band

Black eye

Black
horseshoe
shaped stripe

White chest

Black wings that
they use in the
water like
flippers

Black
chest
speckles

Black webbed
feet with some
pink and black
claws

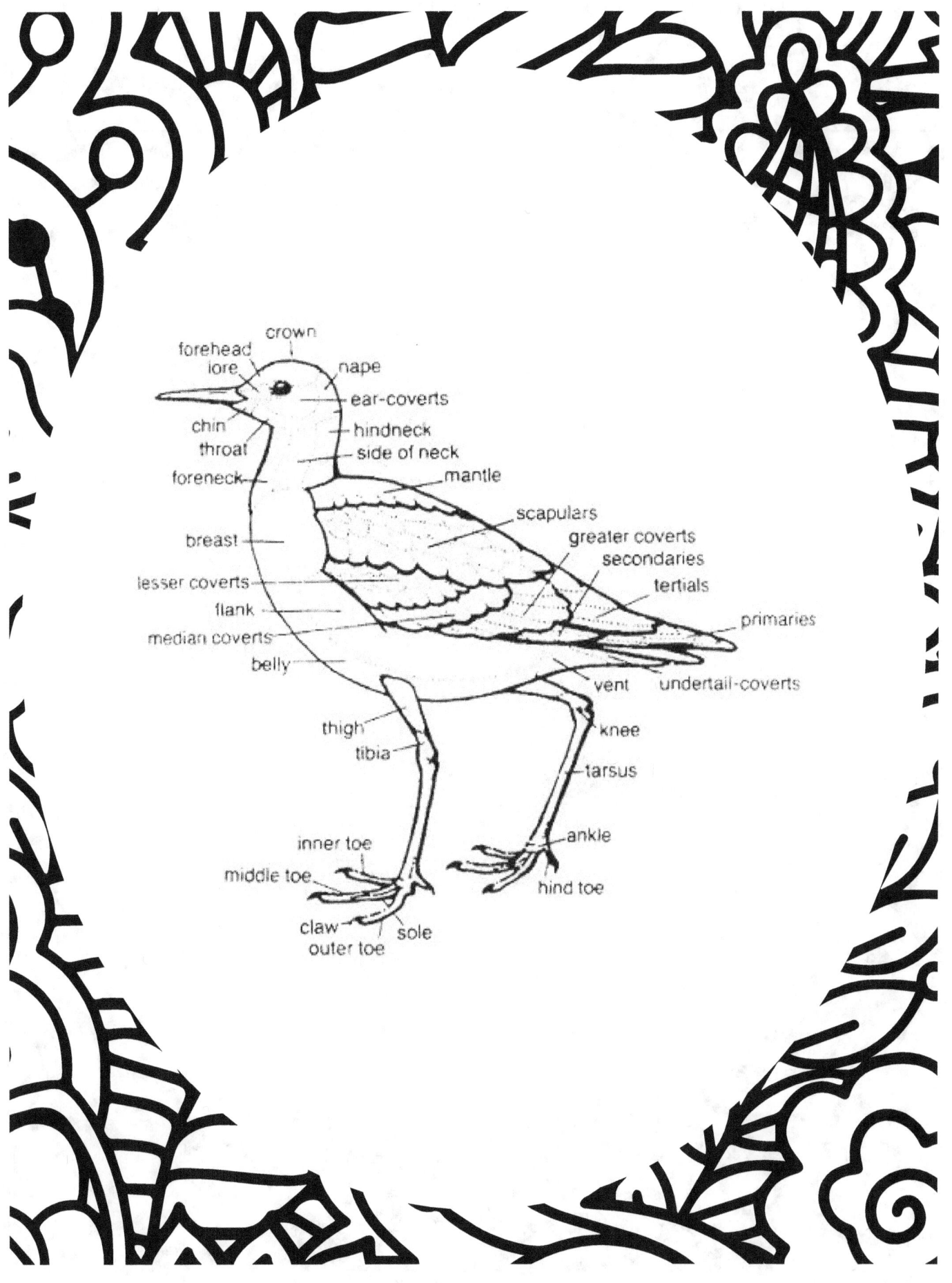

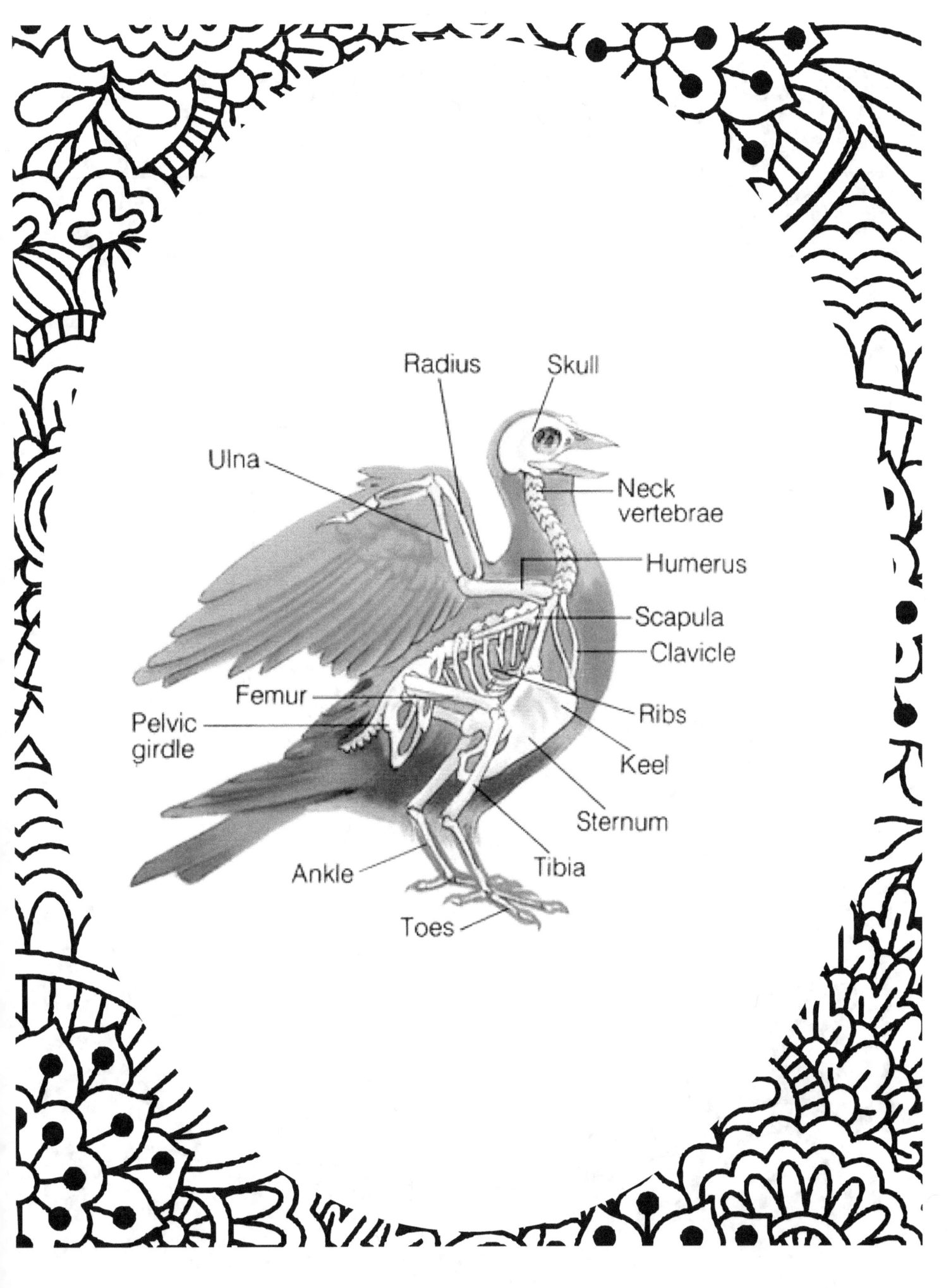

They have a distinctive black cap on top of a pale face and neck.

Males also have a black band across the chest.

The underparts are light colored.

They have long, yellowish legs.

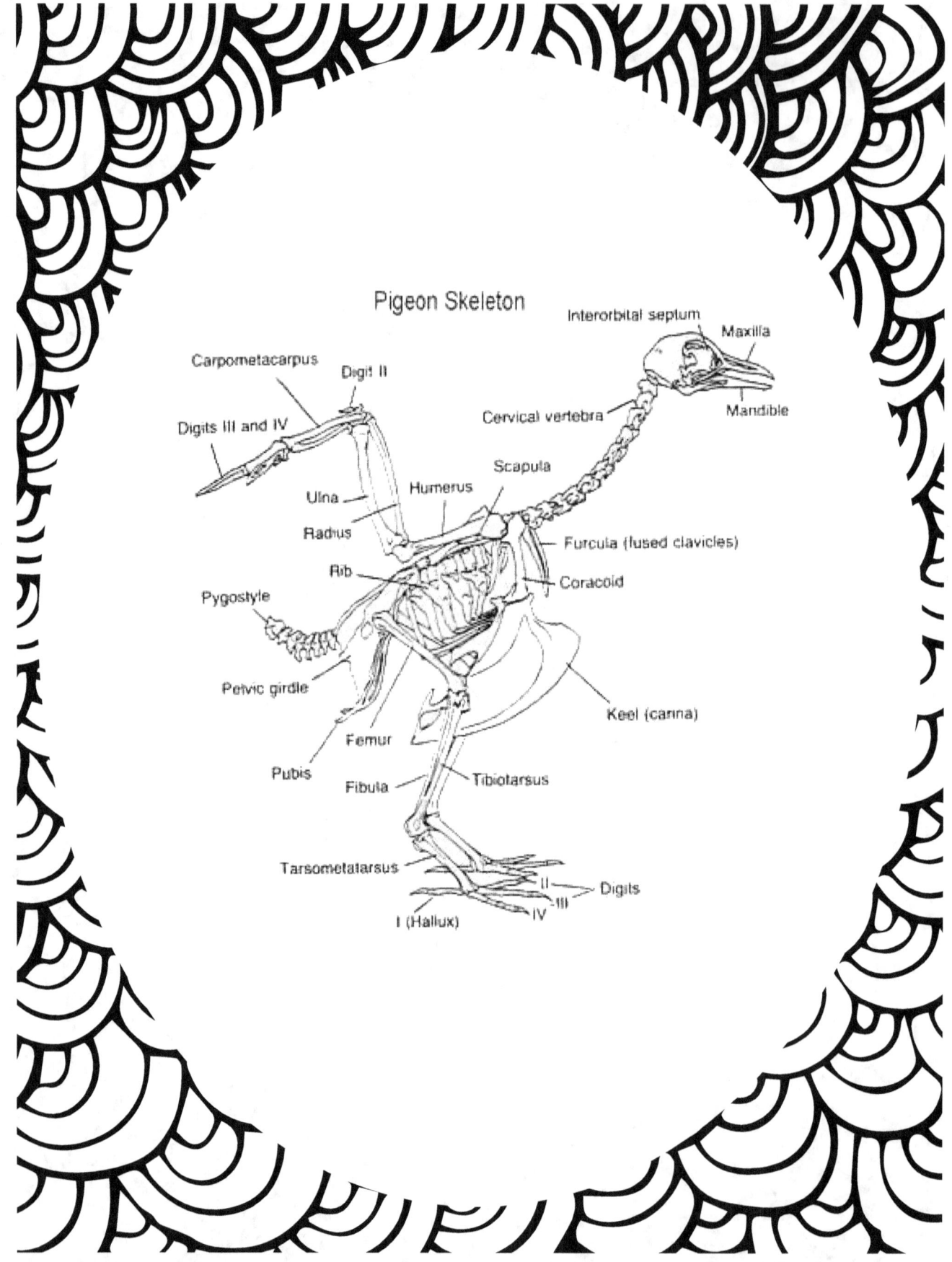

Pigeon Skeleton

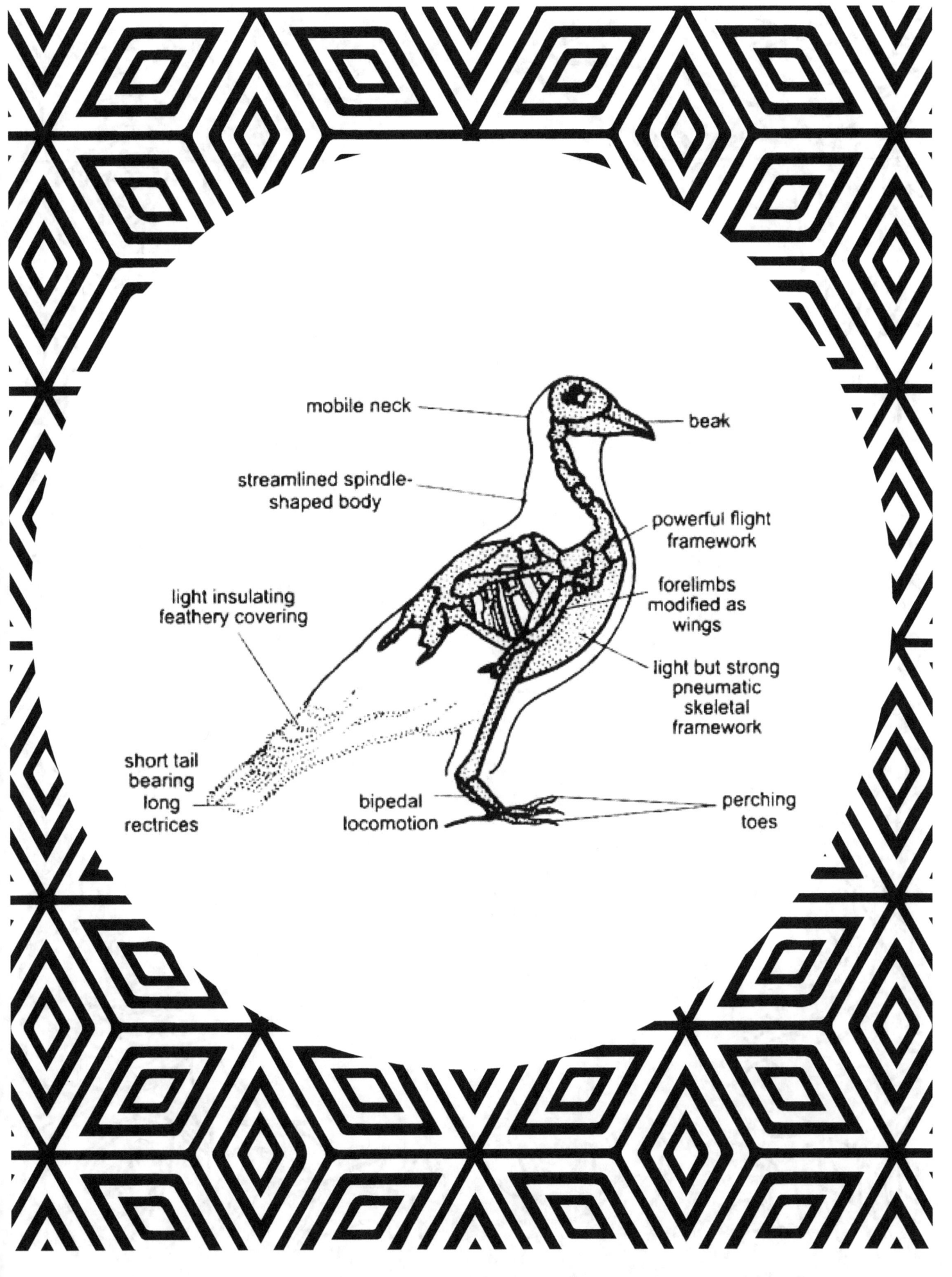

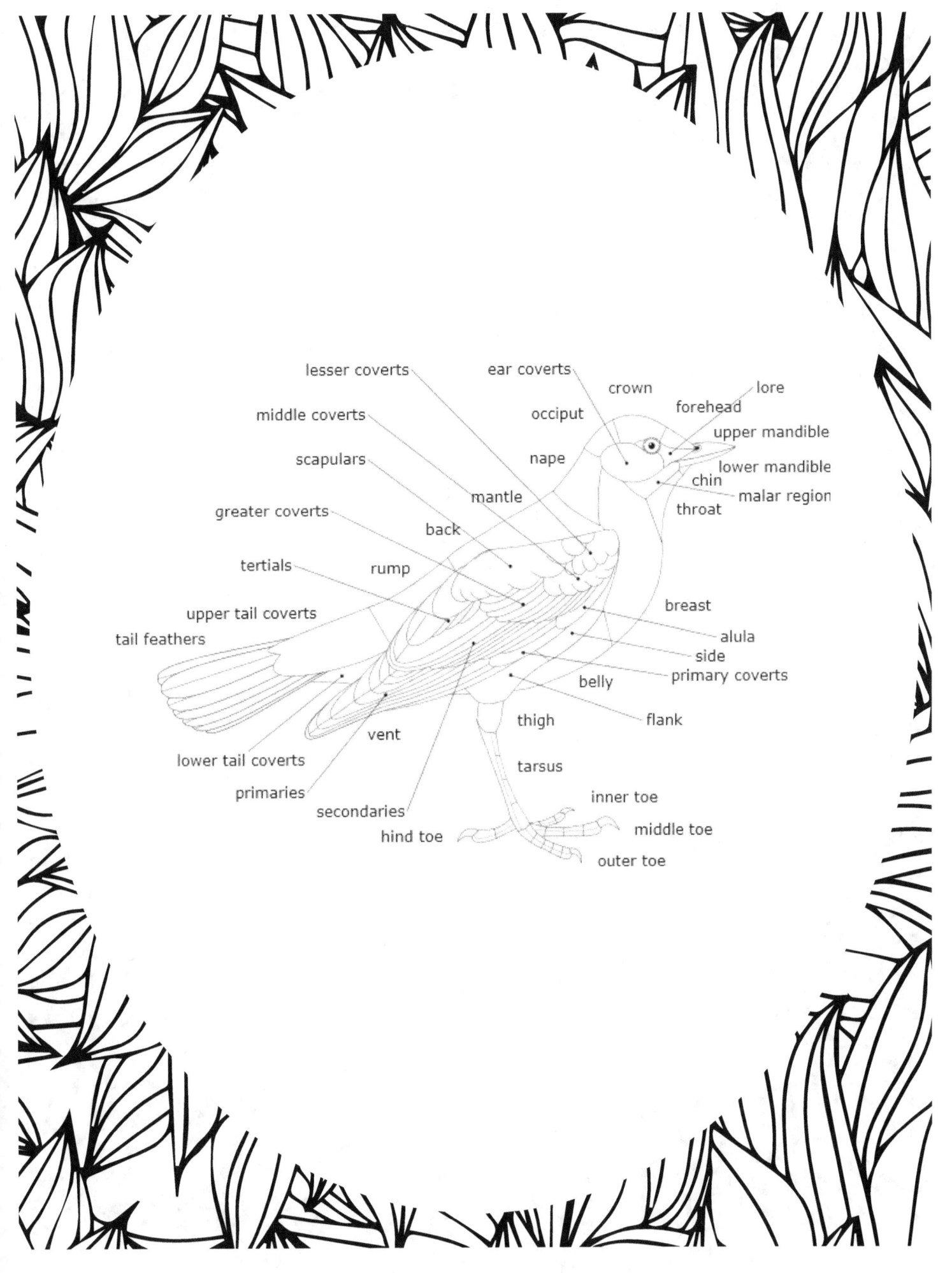

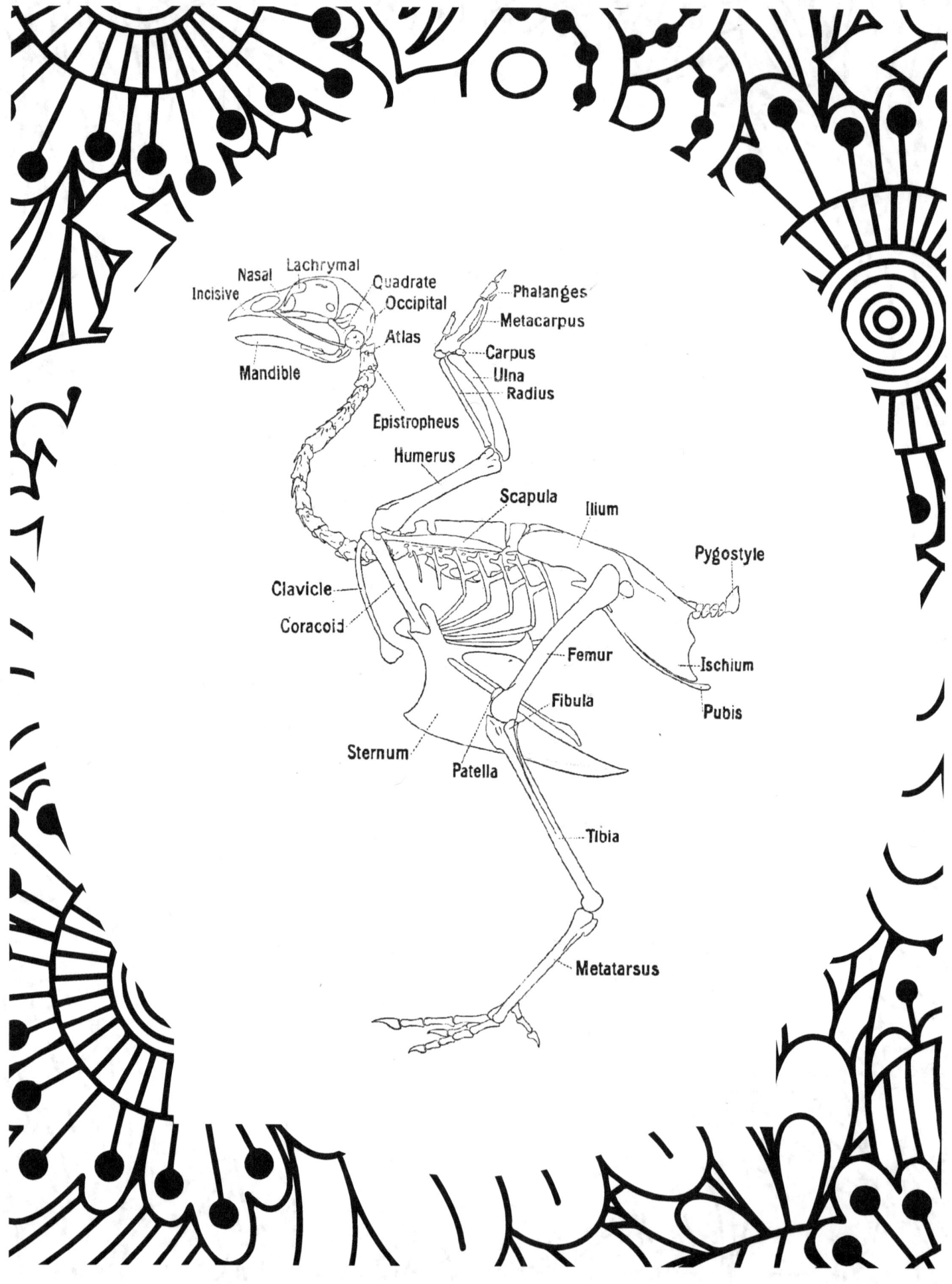

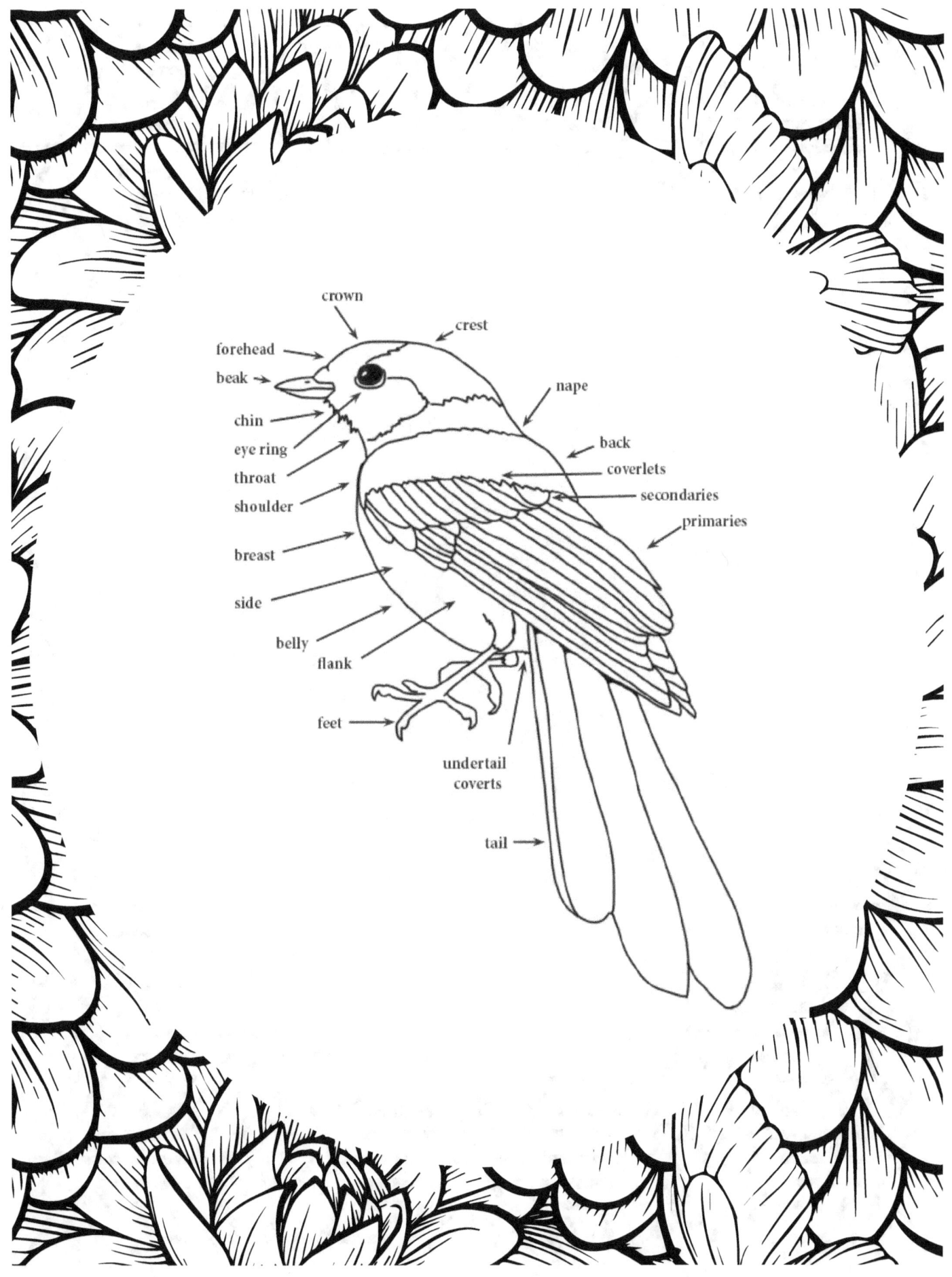

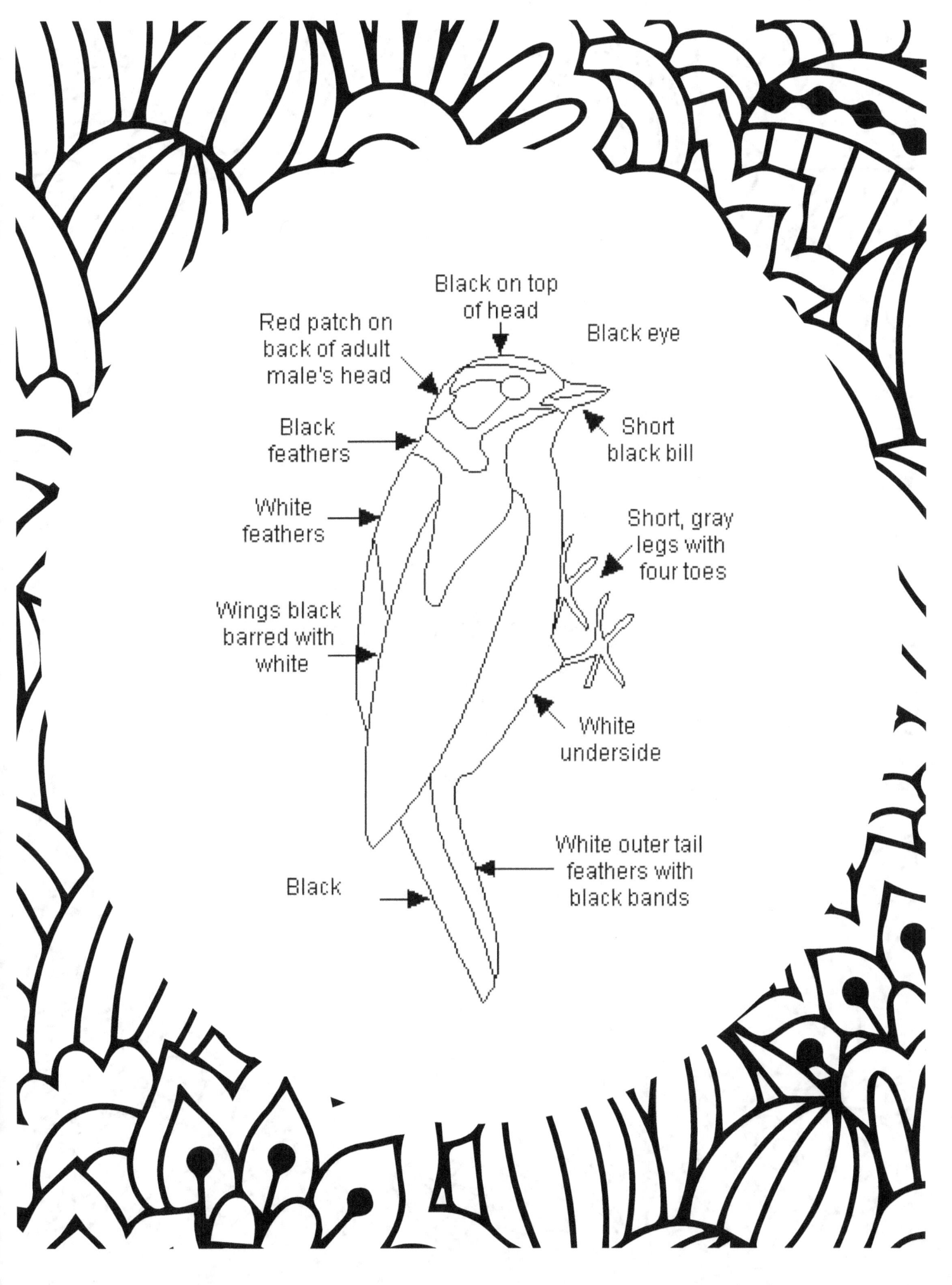

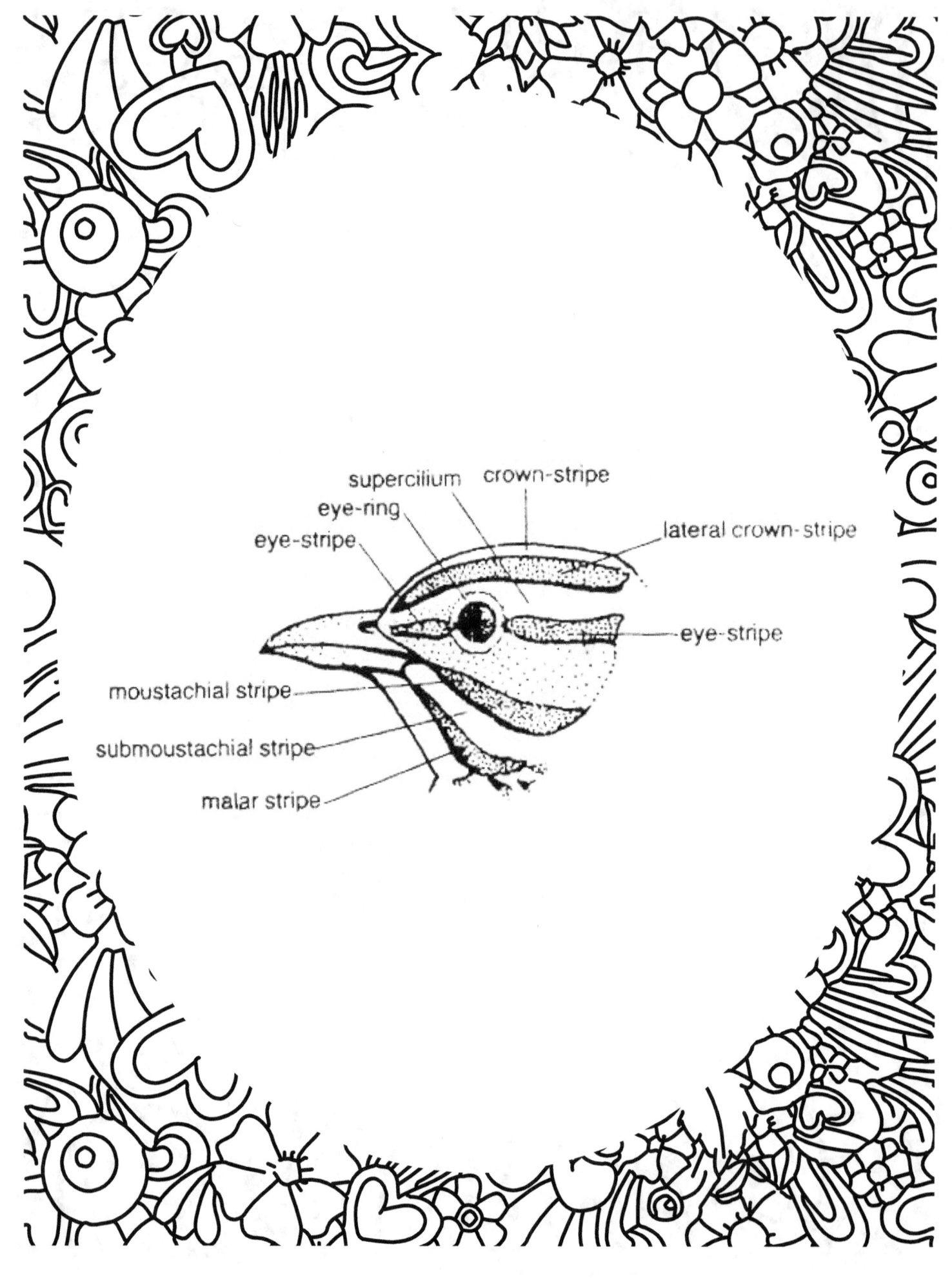

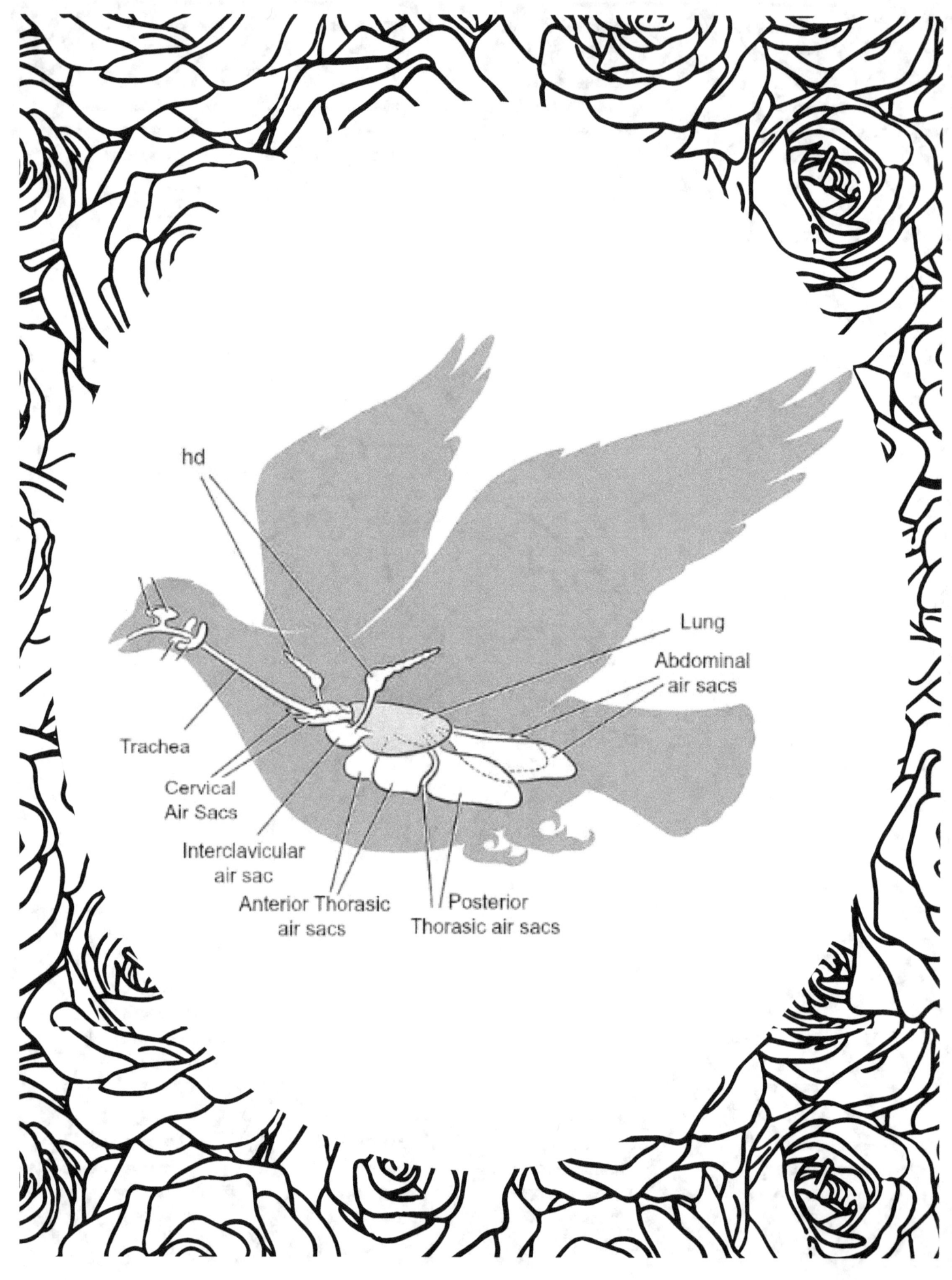

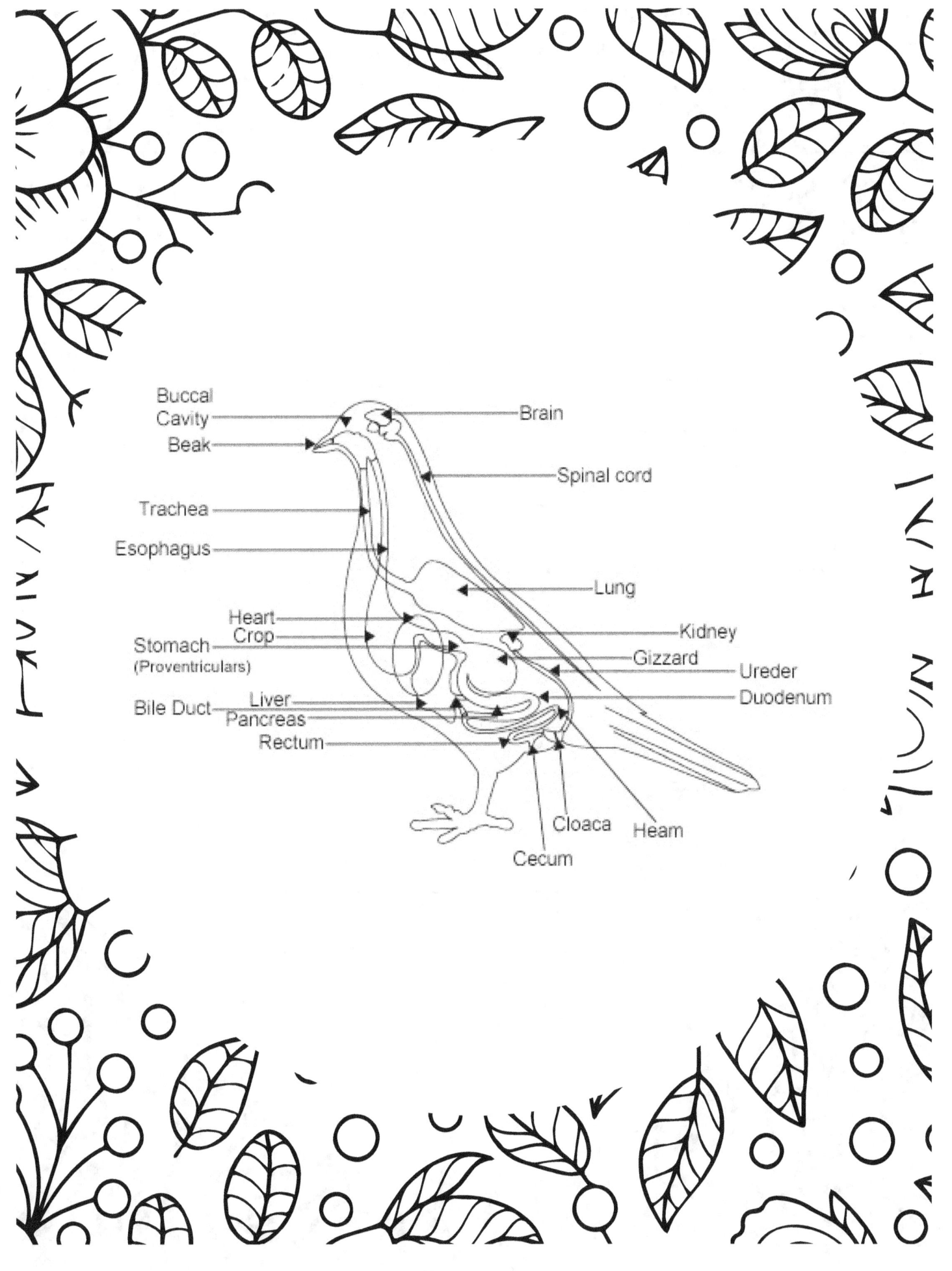

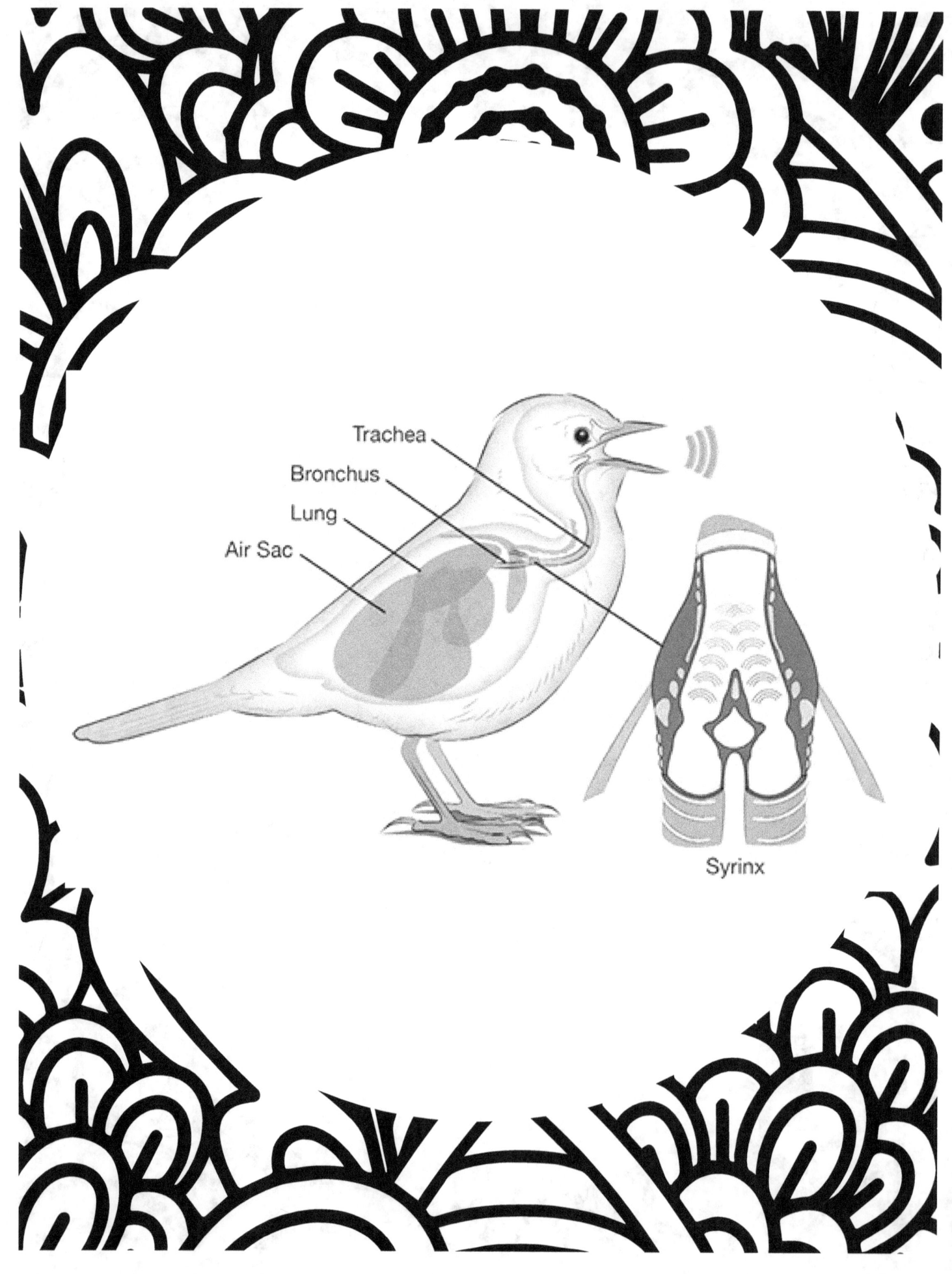

Trachea

Bronchus

Lung

Air Sac

Syrinx

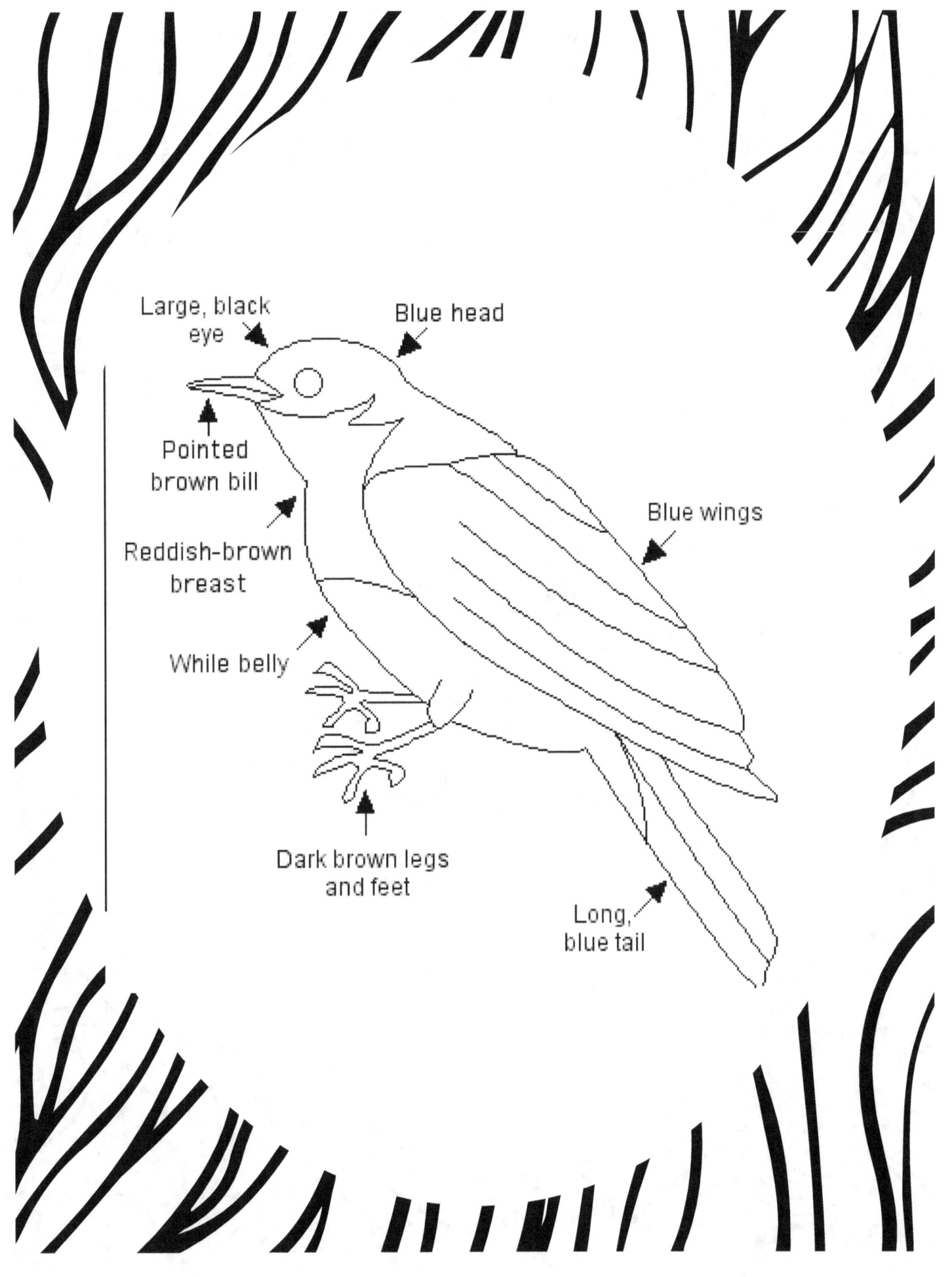

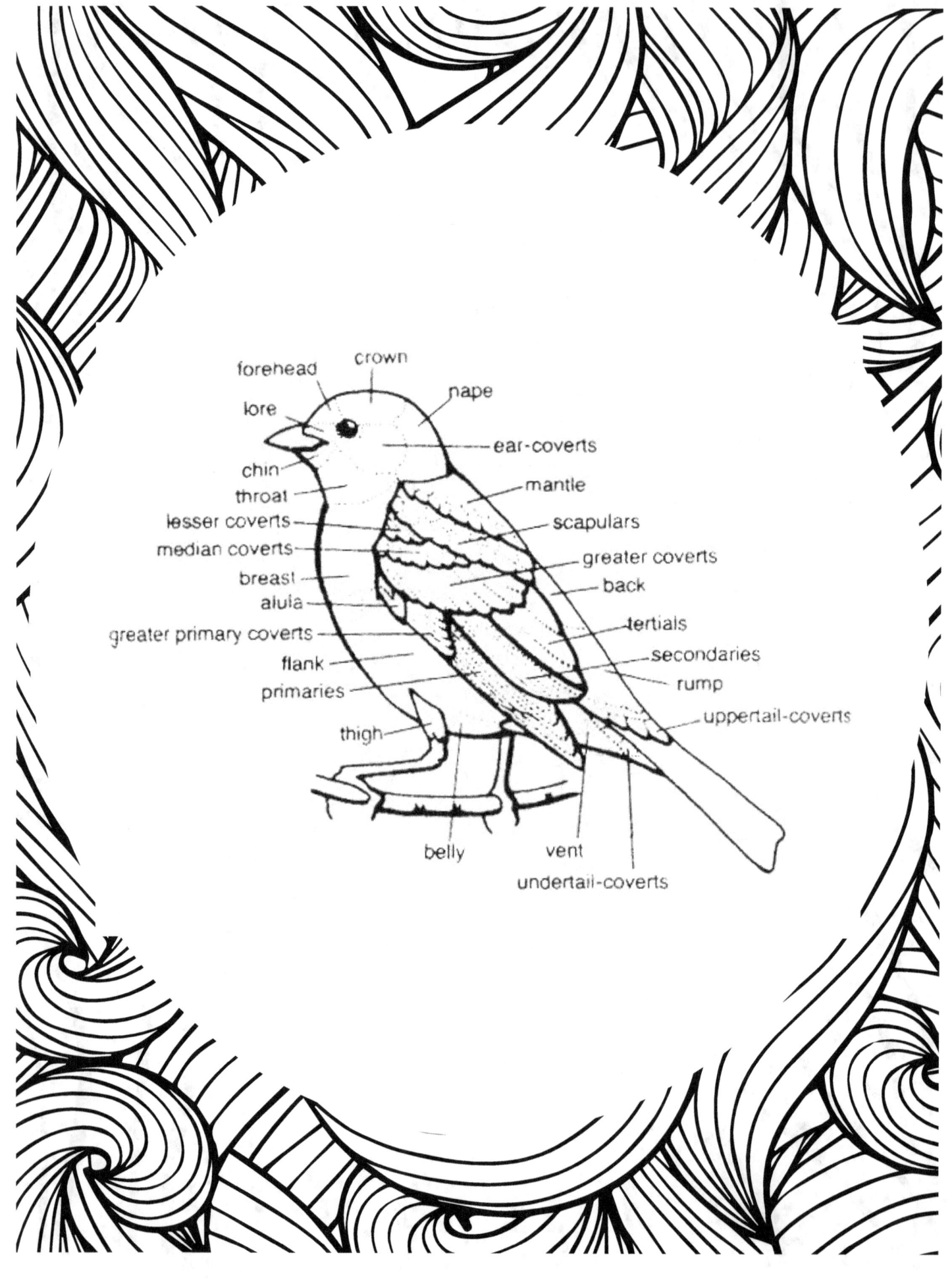

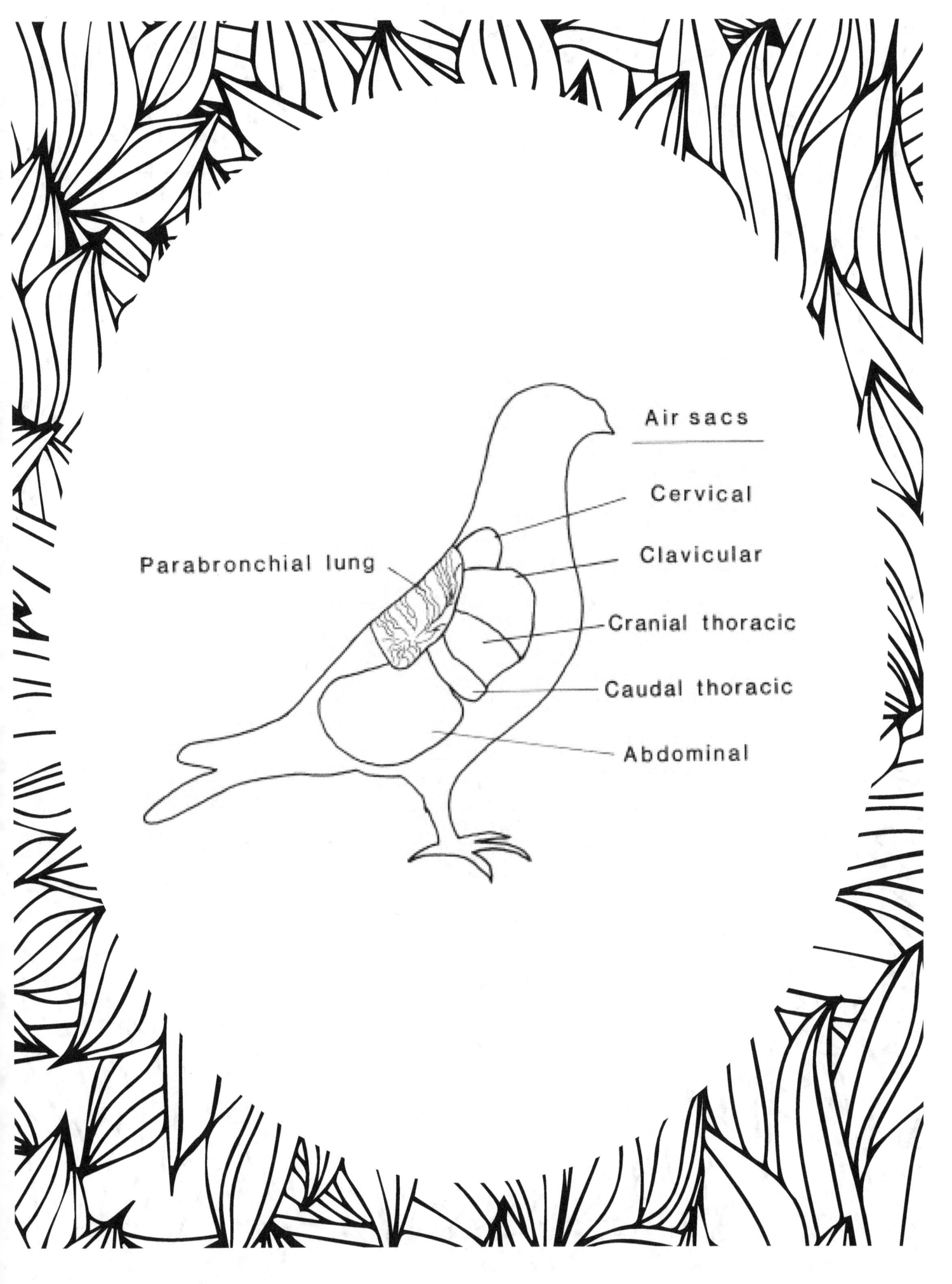

Air sacs

Cervical

Clavicular

Cranial thoracic

Caudal thoracic

Abdominal

Parabronchial lung

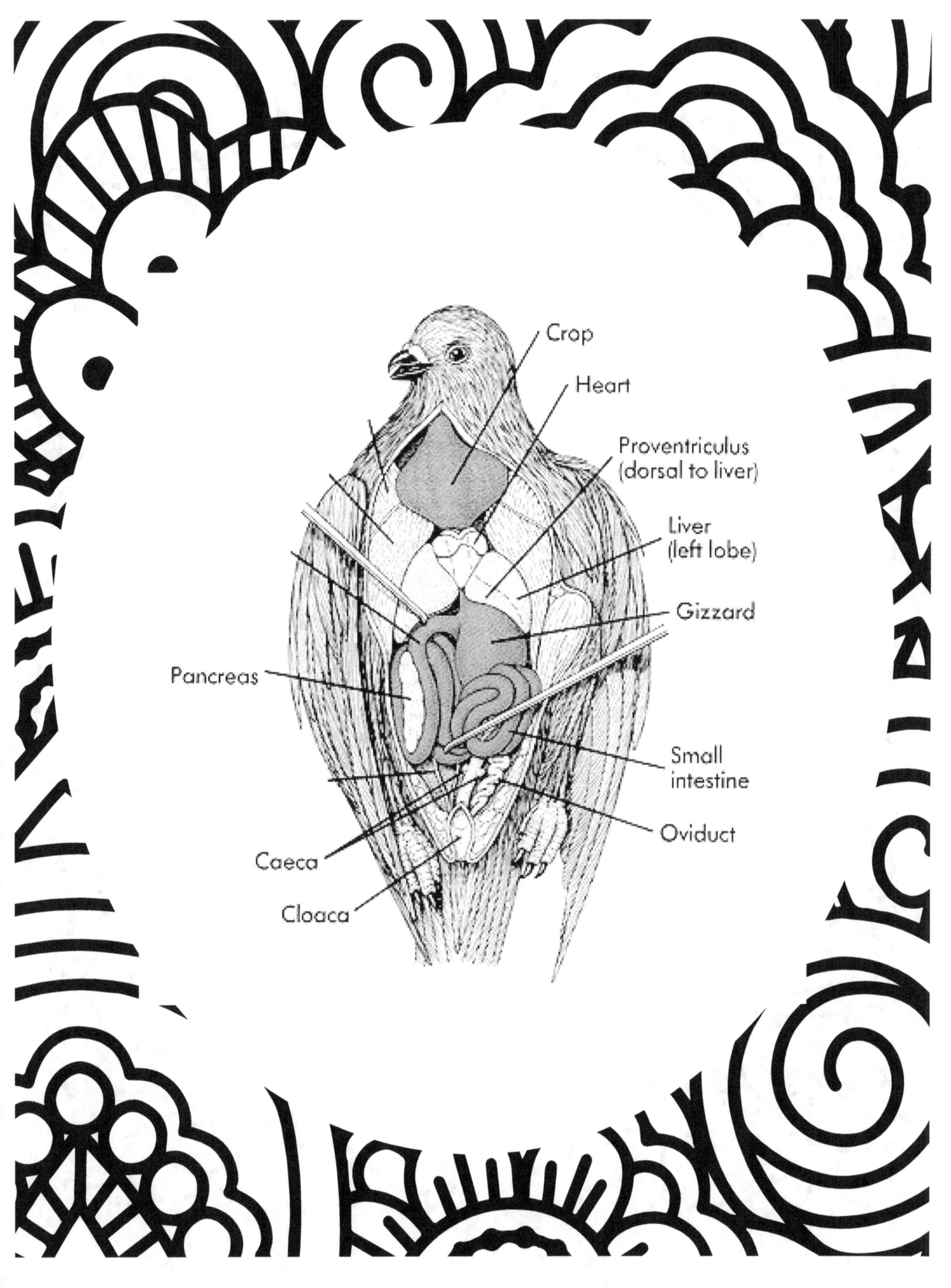

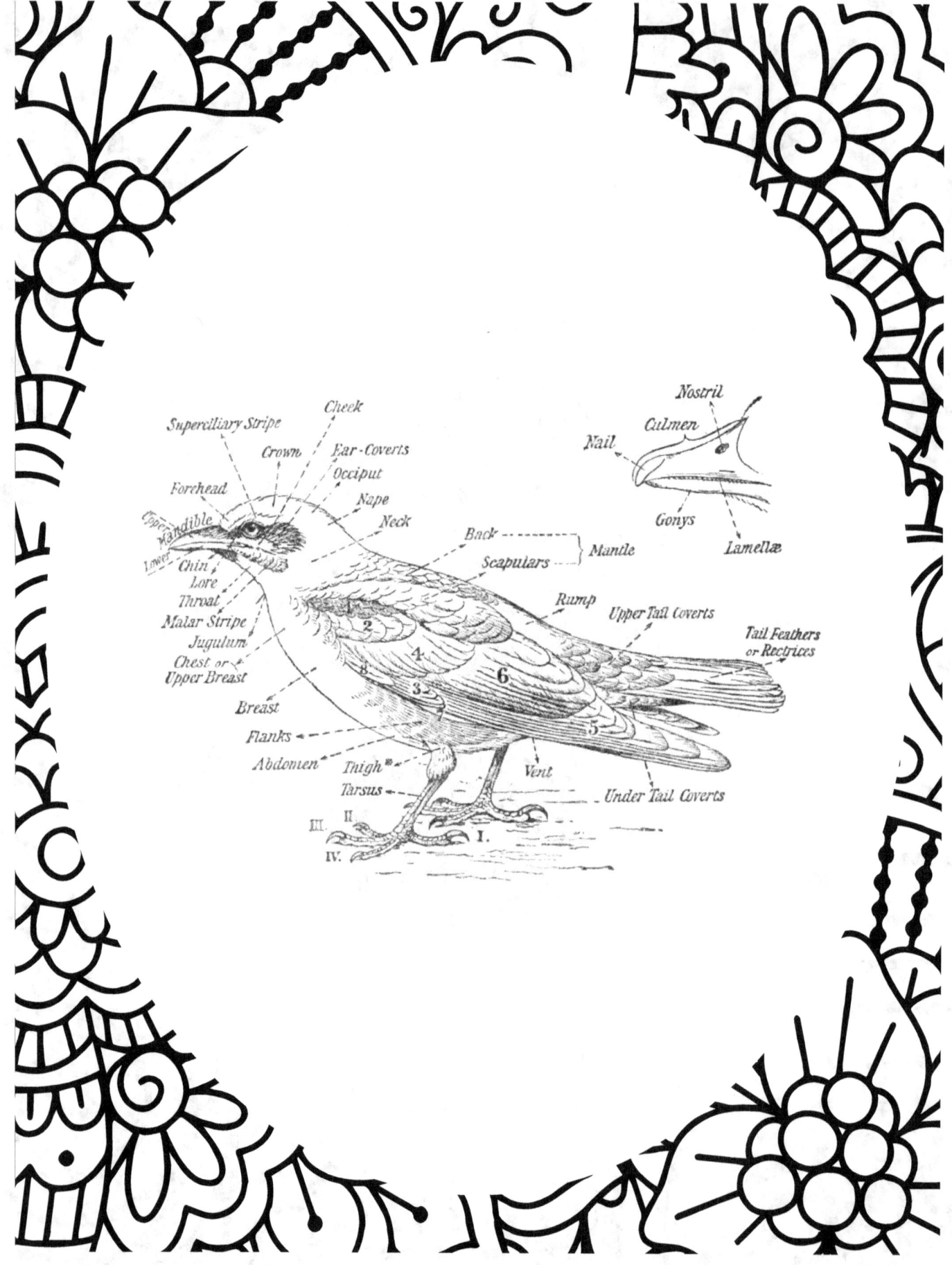

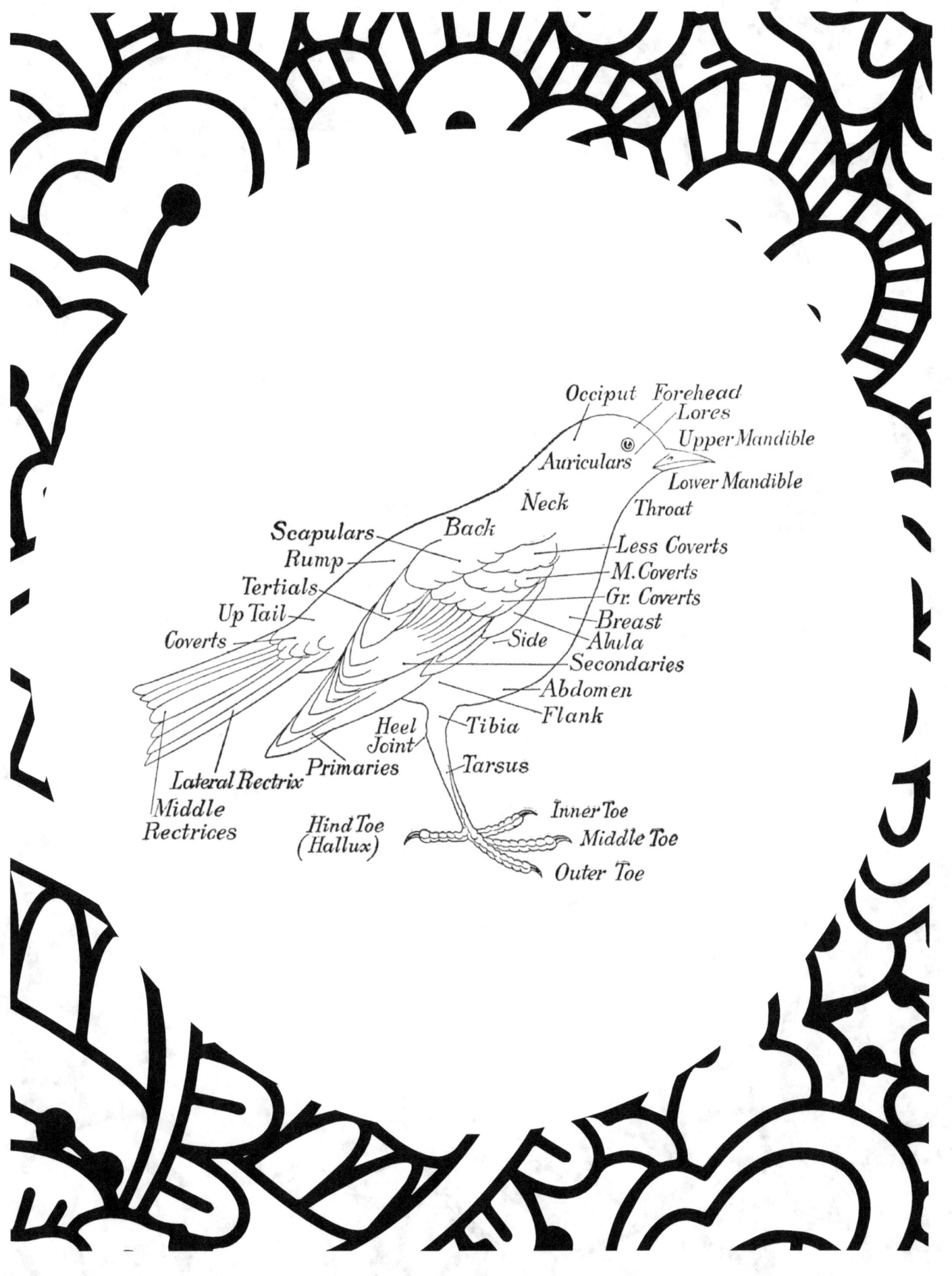

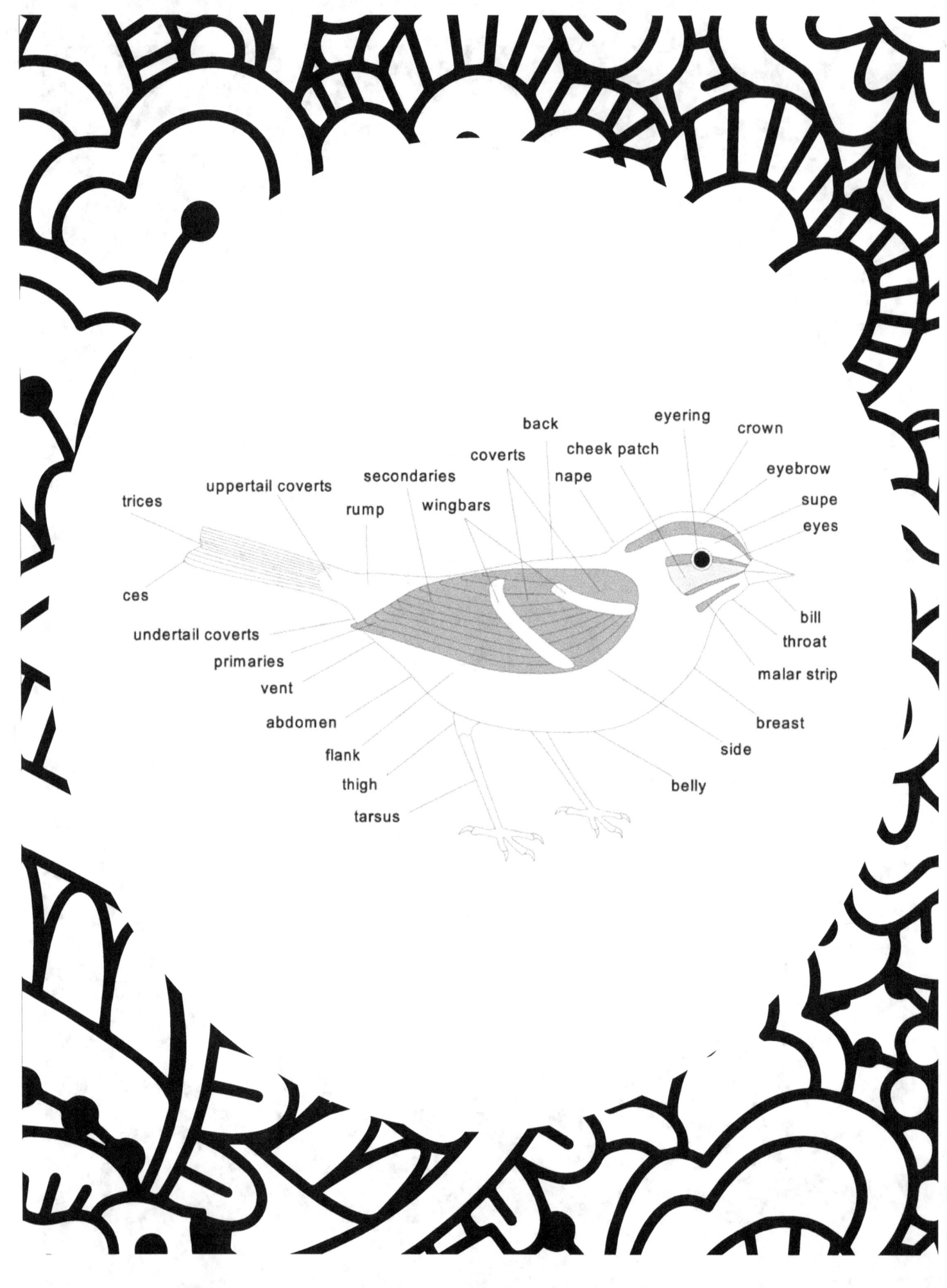